Information Systems Requirements
Determination and Analysis

Donal J. Flynn PhD, CEng, MBCS

Department of Computation
University of Manchester Institute of Science and Technology

McGRAW-HILL BOOK COMPANY

London · New York · St Louis · San Francisco · Auckland · Bogota · Caracas
Lisbon · Madrid · Mexico · Milan · Montreal · New Delhi · Panama · Paris
San Juan · São Paulo · Singapore · Sydney · Tokyo · Toronto

Published by
McGRAW-HILL Book Company Europe
Shoppenhangers Road, Maidenhead, Berkshire, SL6 2QL, England
Telephone 0628 23432
Fax 0628 770224

British Library Cataloguing in Publication Data

Flynn, Donal J.
  Information Systems Requirements:
  Determination and Analysis
  I. Title
  005.74

  ISBN 0–07–707446–7

Library of Congress Cataloging-in-Publication Data

Flynn, Donal J. (Donal James)
    Information systems requirements : determination and analysis /
  Donal J. Flynn.
      p.   cm.
    Includes bibliographical references and index.
    ISBN 0–07–707446–7 :
    1. System analysis.   2. System design.   I. Title.
QA76.9.S88F58   1992   92–798
004.2′1—dc20   CIP

234   HL   943

Typeset by TecSet Limited, Wallington, Surrey
and printed and bound in Great Britain by Hartnolls Ltd., Bodmin.

Cover design by Penpon Ungprateelo

This book is dedicated to Penny, Alex and Clare
for all the love and encouragement they have given to me

CONTENTS

Preface

Overview

A central criticism of Information Systems today is that many systems do not do what their users require, and consequently fall into disuse. A major reason for this is perceived to be a 'computer-centred' emphasis on technical issues in the development process, at the expense of the 'requirements-centred' conceptual issues, related to problem definition and analysis, which are much more important for the correct modelling of user requirements.

The aim of this book is to describe the requirements-centred view, which places the emphasis on the early stages of systems development, where user requirements are determined and analysed, presenting a new, object-oriented, overview model for information systems analysis and describing relevant conceptual issues and models. The application of these to requirements modelling is fully illustrated, using case studies in addition to examples in the text. To reinforce this view, the organizational and social contexts within which information systems are commonly located are also described. The requirements-centred view is basic for an introductory book such as this, as it needs to be developed right from the start of the study of information systems.

A feature of the book is that it addresses the major problems found with information systems and discusses how different approaches provide solutions. This has three advantages:

1. It integrates the book around one theme, as chapters present relevant problems, and discuss how approaches attempt to solve these problems.
2. The problem–solution theme itself is one that serves a reminder of the book's basic message — that information systems should only be introduced where it can be shown that they are a solution to a problem and that technology is not a solution in its own right.
3. It allows the development of a line of argument that, beginning with a statement of problems found with the traditional approach to information systems development, shows, as the book progresses, how modern

approaches to requirements modelling provide solutions to some, although not all, of these problems. This is particularly shown in the contrast between hard and soft approaches to systems development.

The basic message of this book is to convince the reader that information systems should serve a useful purpose in the organization and in the community, and that the use of technology for technology's sake is undesirable as it may have negative effects.

Information systems face three key challenges in the future. Firstly, as has been stated above, many problems are reported with current information systems, and it is necessary for solutions to be found to these problems. Secondly, while information systems have traditionally been limited to processing data for organizational efficiency, future users are going to require such systems to process graphics and speech as well as data. Thirdly, users will also expect information systems to be able to demonstrate, more clearly than before, how they are improving aspects of the organization, such as effectiveness and competitiveness, as these aspects are becoming increasingly important in determining the success or failure of a system.

I use this book as a basis for my Introduction to Information Systems course at UMIST, a first-year course given to BSc Computation, Information Engineering and Joint Honours students. It assumes only an elementary knowledge of programming, the basic hardware components of an information system, and no knowledge whatsoever of the business environment.

Plan of the book

PART ONE: INFORMATION SYSTEMS BASICS

We define an information system and give an example. Various perspectives on information systems are introduced to motivate and explain the background to the area. The main problems found in information systems development are described.

Chapter 1 defines an information system, using a fast-food takeaway example, and develops a general view, emphasizing the organizational context of information systems and the types of activities assisted. The benefits of information systems are contrasted with their problems. Chapter 2 discusses different types of information system and their contents.

PART TWO: UNDERSTANDING INFORMATION SYSTEMS IN ORGANIZATIONS

We develop and describe the OMNIS model, which is an overview model of an information system in its organizational context, and we use this to illustrate the more detailed aspects of information systems.

Chapter 3 introduces many system concepts that are fundamental to the description of information systems, and Chapter 4 builds on these to define the OMNIS model. This will be used later in the requirements determination phase of systems development. A case study is used to illustrate the model. Chapter 5 discusses management activities and information systems, emphasizing the different types of activities and their information needs.

PART THREE: HARD APPROACH TO INFORMATION SYSTEMS

Here the aim is to describe the requirements and analysis phases of the systems development process, showing the early stages of how an information system is built. The chapters on methods and tools show how a common requirement is developed by some well-known methods and tools.

Chapter 6 begins with a description of the traditional approach to the process, discusses problems with the approach and considers alternative approaches, including the iterative, evolutionary, user validation, prototyping and formal methods approaches. In Chapter 7, requirements determination is examined, beginning with problem definition and a feasibility study followed by a requirements acquisition case study. The application to requirements analysis of the OMNIS model developed earlier is then shown with a stock room case study in the Eurobells △ organization.

Chapter 8, Analysis I, illustrates the use of the object-oriented approach to the analysis of entities and rules, using aspects of earlier case study user requirements held in OMNIS. Chapter 9, Analysis II, describes the traditional techniques for analysing processes, and concludes with a description of the object-oriented approach to integrating entity, rule and process.

Chapter 10 discusses some well-known methods used for systems development and Chapter 11 outlines the types of software tools available for method support and describes a particular tool, AUTOMATE-PLUS, in detail.

PART FOUR: SOFT APPROACH TO INFORMATION SYSTEMS

We criticize the basic approach to development described in Part Three by describing the problems that arise when information systems are developed according to this approach, termed the 'hard' approach. Some ideas are put forward for improvements, which are collectively termed the 'soft' approach.

Chapter 12 contrasts the hard and soft approaches, showing how the soft approach addresses problems caused by the assumptions of the hard approach, and Chapter 13 describes an organization theory point of view, considering how information systems can be designed to 'fit' with social factors in the organization. Chapter 14 describes important issues in strategic planning for information systems within organizations and Chapter 15 discusses the impact of information systems on the individual, organizations and society.

Part One
Information Systems Basics

The first part of this book, in Chapter 1, shows how a simple information system works, using the example of the MacAdam fast-food takeaway, and then describes some important features of the system that need to be clearly understood.

Benefits are mentioned and then certain problems with information systems are described. Problems of quality and productivity are discussed in detail and examples are given of how these problems might relate to the MacAdam takeaway. Chapter 2 discusses different types of information systems and shows how they fit the definition given in Chapter 1.

1
What is an information system?

Introduction

Our definition of an information system is:

> An information system provides procedures to record and make available information, concerning part of an organization, to assist organization-related activities.

The aim of an information system is to provide a means for processing information to improve the efficiency and effectiveness of the organization. We will see that this definition falls into two parts: the first emphasizing system structure and functioning, the second concerned with the organizational context of the system.

We shall use the term 'information system' to refer to a computer-based information system, and, in addition, we shall be mainly concerned with commercial information systems, characterized by the fact that people are the main suppliers and receivers of information. These are the type of systems that we find in organizations such as banks, insurance companies, and retail, distribution and manufacturing organizations.

The MacAdam fast-food takeaway

MAIN POINTS

We now present a simple information system which will illustrate the components of our definition. Figure 1.1 shows an information system that is used in a branch of the MacAdam fast-food takeaway, selling food such as beefburgers, French fries and soft drinks. The system has been in full use for six months and all staff report general satisfaction with its operation, finding

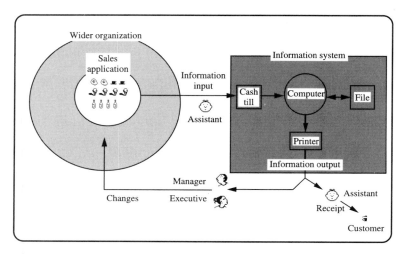

Figure 1.1   Information system in the MacAdam takeaway

that it helps them to do their jobs better, that it is easy to use and that system reliability is good, with few problems.

The main points of the system are:

1.   When items are sold, a sales assistant enters information into a cash till connected to the computer. An entry consists of data identifying the item and the quantity, for example, six chocolate milk shakes.
2.   The system knows the price of each item and calculates the total price.
3.   Each entry is stored on a computer file.
4.   Information concerning the sale is produced by the system, in the form of a printed receipt, which is given to the customer by the sales assistant.
5.   Sales information is available for the takeaway manager, John Silver, who can obtain a printed report from the computer file showing, for example, sales yesterday or this month, and numbers of customers.
6.   Information is also available for Sally Cuthbertson, the sales director, who is based at Head Office in Glasgow, consisting of quarterly sales to date, number of customers weekly and quarterly, and sales for each product.

   In addition, an analysis program can present Sally with predictions of the profiles of proposed products, in terms of business factors such as likely market share and short-term (1 to 6 months) and medium-term (6 months to 2 years) earnings, based on existing product consumption.

STRUCTURE AND FUNCTIONING OF THE INFORMATION SYSTEM

The basic features of the information system in Fig. 1.1, in relation to the first part of our definition, are as follows. First of all, *information* is recorded in the system, on a computer file. The information *concerns a part of the organization* as it relates to the sales function, recording information such as items sold, their quantities and their price. There are *procedures to record and make available* the information, as the assistant is helped to enter information into the computer and receives a printed receipt, and managers and executives can obtain reports that present the recorded sales information in different ways.

The human procedures related to recording and making information available initiate procedures carried out by programs (software), such as price finding, item totalling, data storage and report production, which in turn control hardware devices (computer, cash till, printer and file).

The information system in this example supports the sales function of the takeaway, and we term this a *sales application*. There may be other applications, for example the purchase of food from suppliers.

When items are sold, each sale constitutes a *transaction*, which is a basic, everyday process in the organization. Another example of a transaction might be the delivery of a quantity of food, such as a thousand frozen beefburgers, from a supplier.

ORGANIZATIONAL CONTEXT OF THE INFORMATION SYSTEM

Turning our attention to the second part of the definition, we want to emphasize that the information system is used *to assist organization-related activities*. This means that there is a close relationship between the information system and the organization of which it is a part.

This relationship is characterized by the fact that, firstly, the information system satisfies the *information needs* of the required activities. Secondly, it provides automated *procedures* which assist or replace some activities. Thirdly, the system is *usable* by and *acceptable* to the organization.

Information needs

- *Sales assistants*. The sales assistants are provided with item prices by the system, which also provides information concerning the sale on a receipt to give to customers.
- *Takeaway manager*. John uses information from the system to check that the takeaway is performing to expectation. There are targets for numbers of customers, for sales and for staff costs, and if targets are not being achieved then he must take decisions to reduce costs or lower profits on certain items to stimulate more business. In this way he uses the system to control the business.
- *Sales director*. Sally uses information from the system to help her decide if any changes are required to the products sold by the takeaway. For example, pomegranate-flavoured milk shake may be discontinued if its

sales are down for six successive months. Using the basic sales information, the system can accumulate and analyse the figures to present the information she requires to take such product decisions.

Procedures

- *Sales assistants*. Procedures are provided for recording information that allow sales details to be entered, check that the details are valid, calculate total prices and store the entries on the file. Procedures for making information available are also provided, as the system prints the sales receipt automatically, so the assistants do not have to take time to write it out manually.
- *Manager and director*. For John and Sally, the relevant procedures produce the information mentioned above to assist their activities.

Usability and acceptability

The third characteristic of the relationship between the information system and the organization is that the information system is *acceptable* to the users, as they report that they need the information from the system and feel that it helps them to organize their jobs better. In addition, the information system is *usable*, in the sense that the system is not hard to operate, it is reliable and it allows information to be input and output in as simple a fashion as possible.

INFORMATION SYSTEMS DEVELOPMENT

The development of the information system for MacAdam's followed the traditional life-cycle path. This generally consists of several phases, as follows:

1. *Requirements*. The requirements for the system are obtained from the users. These set out what the system should do and how it should fit in to the organization.
2. *Analysis*. The requirements are analysed and a detailed specification is produced.
3. *Design*. This is often separated into logical and physical design, and it is the phase where the detail of the system is worked out, including computer and manual processes.
4. *Implementation*. Programs are written, the system is tested and it is put into operation.
5. *Maintenance*. Users request changes to the system which are analysed, designed and implemented.

The MacAdam sales information system was developed over a period of twelve months by the takeaway Head Office, underwent a pilot test in several selected branches and is now installed in all branches.

Types of activities and information

We now present a more general view of an information system, focusing on the main types of organizational activities, their information needs and related procedures. Figure 1.2 shows the context implied by the above discussion, where an information system provides automated procedures and information to assist organization-related activities for an application.

ACTIVITIES

The activities in an organization are usually classified into two broad types:

1. Functional activities.
2. Management activities.

Functional activities are performed by junior employees in an organization, such as the sales assistants in MacAdam's. These activities transform different types of input, such as raw materials, money and information, into the goods or services of the organization, as well as carrying out related tasks such as processing customer orders, accepting payments and processing organizational information. In contrast, *management activities* are performed by different levels of management, from supervisors to senior executives, such as John and Sally at MacAdam's. Management activities plan and control the organization and direct its functional activities.

A distinction that may be drawn between these two broad types of activities is the degree of structuring of the decisions that may be made. Management have considerable freedom in the decisions they make, while junior employees can only make very restricted types of decision. For example, a sales assistant in MacAdam's might not be allowed to accept a cheque in payment without authorization from the supervisor.

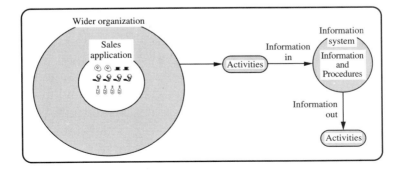

Figure 1.2 A general view of an information system in its organizational context

A consequence of this is that many functional activities, being more routine, are more easily automated, and hence can be partially replaced by information system procedures.

INFORMATION NEEDS, ACTIVITIES AND PROCEDURES

We shall briefly look at typical management and functional activities and consider their information needs which are satisfied, as well as the activities that are assisted or replaced, by an information system.

Functional activities

Some activities, termed information processing activities, are concerned both with recording information as well as receiving information from the information system. They may also receive information during the input process. For example:

- *Processing a shipment order.* The customer name and order number is entered; the system checks the order number, supplies the customer address and prints the despatch note. The shipping clerk sends the despatch note with the order.
- *Paying employee wages.* The employee name and code is entered and the system checks the code, supplies information concerning the rate of pay, calculates the total pay and prints the pay slip. The wages clerk gives the pay slip to the employee with the wages.
- *Processing a MacAdam sale.* The sale details are entered and the system supplies the price and calculates the total. The amount tendered by the customer is entered, the change is displayed and the receipt is printed. The assistant gives the customer the receipt and change with the customer's food.

Other types of activities may only use information from the information system and will not input information. For example:

- *Locate warehouse part.* The part code is entered and the system checks to see if there are any in stock. If there are, it displays the location of the part together with the quantity remaining. If out of stock, it displays an appropriate message.

  We may include under this category those activities that only disseminate information to other employees.

A characteristic of functional activities is that they are typically a mixture of computer procedures and human activities.

nuclear power stations (for example, Chernobyl), railways, aircraft and missile early warning systems. Recent cases involving banking systems have involved incorrect transfers of hundreds of millions of pounds.
5. *Security*. Reports are increasing of computer fraud, where information is accessed that is available over networks vulnerable to intervention by unauthorized individuals (such as hackers). There is also a growing problem caused by illegally introduced software (such as viruses) which corrupts systems.
6. *Big Brother*. This problem concerns the fear that information concerning individuals may be used for personal surveillance by government or related bodies.

PROBLEM REPORTS

Many information systems are never delivered or are never used, with a figure in the region of 40 per cent being suggested (Eason, 1988). In addition, perhaps only 20 per cent of systems have a positive effect on organizations, the remaining 40 per cent having only a neutral effect.

A recent UK survey by the consultants KPMG, Peat Marwick McLintock (KPMG, 1990) found that '30% of the UK's biggest computer projects were massively over budget, over time, and, if ever completed, fail to do the job they were meant to'. It is suspected that unsuccessful systems may also lose customers and eventually threaten organizational survival. For 'front-office' applications, a 1986 survey by the DTI in the UK (DTI, 1986) showed that, on average, of twenty pilot office automation projects, only half worked as the users expected, were fully accepted or brought positive benefits to the organization, and 20 per cent were rejected.

An analysis (Humphrey, 1988) by the Software Engineering Institute, based at Carnegie-Mellon University in the United States, evaluated approximately 150 of the leading software producers in the United States and found that over 80 per cent of producers had a development process considered to be 'chaotic', with problems in the areas of progress planning and change control.

Business reports in the United States (McFarlan, 1981) indicate major quality and productivity problems with commercial systems, and in the United Kingdom the literature contains similar problem reports (Fawthrop, 1990; Fairbairn, 1989).

QUALITY

Reasons for poor quality

It is, unfortunately, only too easy to develop an information system that does not have a beneficial effect on the organization. This may occur for several reasons:

1. The information system may address the wrong problem, in that it may not help (or may help only marginally) the organization become more efficient or effective. There may be a more important problem that is missed. Alternatively, the information system may conflict with organizational aims. In many cases, the wrong activities to assist are chosen.
2. Wider social or psychological factors may be neglected, such as the existing reporting structure of the organization or the extent to which the information system will be acceptable to or usable by its intended users.
3. Even if the right activities are identified, the organization may be incorrectly analysed for its information needs, due to organizational or technical ignorance on the part of users or analysts.
4. The system may be developed for the wrong reasons, such as technology push from technical experts or political pull from ambitious managers.

The majority of these problems are due to a lack of attention to the nature of the relationship between the information system and the organization.

Existing MacAdam system

In the takeaway, the users report that the system has a beneficial effect, so it may be assumed that the problem solved by their system is the correct problem. Firstly, the activities identified, which are concerned with controlling costs and identifying slow-selling products, are activities that are agreed upon as being important for the success of the organization. Secondly, users are happy with the integration of the system into the organization. Thirdly, the activities have had their information needs correctly analysed and, fourthly, the cost and sophistication of the technology chosen for the system is commensurate with the benefits expected.

Overall, the performance of the activities is improved by computerized provision of the specified information.

Poor quality MacAdam system

However, MacAdam might not have been so fortunate in the quality of its information system, and we may now imagine the sort of problems that might have occurred.

• *Wrong problem.* This problem might have been caused by the selection of management activities that were not important for organizational success. Such a selection might have been made due to a lack of understanding of the organization. A more important activity might have been, for example, to closely monitor the age and gender profile of customers to be able to offer new products that are suitable to the different groups. Alternatively, it might have been better to decentralize decision making from Sally to John for new products, as he is on the spot and can respond more quickly. Another possibility is that wider organizational strategy may emphasize the goal of increased product sales through improved product quality, rather than extending the product range. Hence, the information system would be in conflict with this strategy.

It is probably clear from the above that the choice of organizational activities to be supported by an information system may not be obvious and may depend on intuitive business knowledge. If a mistake is made, then an expensive investment is wasted.

- *Neglect of wider organization.* A problem of this type might have been caused if the existing system had not been piloted in some of the branches. As a result of comments from the pilot, changes were made to the system interface, and a training programme was established in every branch to discuss features and the potential for organizational improvements resulting from the system. Had this not been done, users might have been reluctant to fully use the system, especially as it was devised and imposed by Head Office, and its acceptability would have been low. Such systems often fall into disuse.
- *Incorrect analysis.* The information needs might have been incorrectly analysed, so that sales were only shown by category of product, for example by milk shake, instead of for each type of milk shake. Reports might therefore be useless, and delay would occur while a change request was examined for its implications.
- *Wrong reasons.* An example of this problem might have occurred if the MacAdam technical experts had been allowed to acquire the latest type of database for the system. It would probably have been unsuitable, as it was more expensive than the industry standard, and it was a new product which had not been tested properly and might continually break down, threatening system acceptability and cost.

Solutions

All solutions concern modifications to the basic systems development process. The first problem is addressed by several different approaches, including the soft systems and strategic planning approaches, discussed in Chapters 12 and 14, which aim to identify the right problem to be solved.

Solutions for the second type of problem employ an organization theory approach, discussed in Chapter 13, which with the human factors approach, described in Chapter 12, aims to involve the users more actively as participants in the development process, including acceptability and usability factors as part of the system requirements.

To address the third problem, there is a growth of user-oriented modelling and requirements engineering methods, such as the OMNIS model discussed in Chapters 4 and 7, although most of these require domain knowledge concerning the application to increase their effectiveness.

There is no specific approach for the fourth problem, although group sessions are often used in requirements acquisition (discussed in Chapter 7) to identify and resolve disagreements between users over requirements.

PRODUCTIVITY

Cost overruns of several times the original estimates for development are not uncommon. Such overruns will affect cost/benefit justifications made for the system. If the estimated delivery time extends from 18 months to three years, for example, it is possible that the system may not be useful, as events in the market may move so quickly that it is out of date when delivered. Both these factors may cause systems to be cancelled or never used, once delivered.

Although some of these problems are due to poor estimation techniques and project management, the most important factor is changing requirements. It is a well-known phenomenon that requirements for an information system may be in a state of flux throughout the systems development process. If the specification of a system is always changing in this way then new work must be done and completed work redone, both factors causing costs to grow and delivery times to lengthen. The problem of changing requirements takes place within the context of a shortage of experienced development staff.

Reasons for changing requirements

There are three main reasons why requirements change:

1. Users often have only a vague notion of requirements at the beginning of a project, learning them more thoroughly as the project progresses.
2. Changes in external factors such as technology, legislation and the market will change requirements for systems.
3. Requirements may have implications for implementation which are not feasible, but which are realized only at implementation time.

Problems of quality and productivity will interact, as, for example, the discovery during testing that a system is of poor quality may mean a delay while the system is put right.

Existing MacAdam system

The way in which the information system was developed at MacAdam was only discussed briefly. However, the fact that the system is operating successfully does not mean that it met its cost or time development targets. In fact, as it is difficult to estimate the benefits a system brings, it is only rarely that it is possible to measure whether a system is actually of financial benefit to an organization.

minimize travel costs by scheduling deliveries more efficiently, and economic forecasting, where information concerning exports, imports, interest rates and so on is input to a model of, for example, the British economy.

Another example is a share-dealing system where the computer analyses stocks and their prices against expert criteria, but must present a list for a final decision to a human, who uses intuition and 'seat-of-the-pants' expertise. Not all of the human processes may be precisely specified in this system.

Typically, such systems are information systems for individuals or small groups in an organization, used to plan for the future. In contrast, transaction processing systems are designed to monitor daily operations. DSS/EIS are often difficult to design as they need to be closely integrated with decision-making activities of managers.

REAL-TIME SYSTEMS

Real-time information systems are commonly met in organizations specializing in defence, producing aircraft, tanks and submarines; embedded processing, producing washing machines, car electronics and security alarms; and industrial control systems, concerned with power stations, refineries and so on. Some computer systems used in real-time systems form part of a product that is being made, and most of their inputs and outputs are typically in non-character form and are transmitted to and from control devices, as opposed to people. This contrasts with the commercial information system, which stores and makes available character-based information, almost exclusively to people, concerning the product or service of the organization; that is, it is not part of the delivered product or service. However, a real-time system in a chemical plant, a nuclear power station or a railway network will invariably provide information for individuals to check system functioning.

The use of the term 'real time' probably arose in the early days of commercial information systems, which did not operate in real time but which used input devices such as punched cards and magnetic tape for information storage, both of which caused delay in capturing, processing and presenting information. The hallmark of a real-time system is that the computer must be able to respond to input signals immediately. For example, if a car wheel sensor signals an approaching skid to an automatic braking system, a five second delay may mean the system will not survive!

A real-time system will usually combine a transaction processing system with a decision support system for analysing the basic data.

DATABASE SYSTEMS

Most organizations contain information that is used by more than one application, for example information concerning departments and employees that is required by a personnel as well as a project management application. Traditionally, each system would have duplicated some or all of this information on its own files. The trend is for applications that use large amounts of shared information to use *database systems*.

These contain an integrated collection of files (the database) for recording the information, as well as software for providing fast access, often with a user-oriented interface. They are integrated as they remove the redundancy that often occurs when separate files exist, as well as containing links between related items, for example the links between departments and employees that record the department in which an employee works. Typically, they are used to store a large amount of functional data centrally, for sharing by several transaction processing systems, so that users in the different systems can access the data at the same time.

EXPERT OR KNOWLEDGE-BASED SYSTEMS

It is rather more difficult to distinguish knowledge-based applications from the systems discussed above, as they are a more recent development and general agreement does not exist on their contents. However, *knowledge-based systems* (or *expert systems*) share similarities with decision support systems, in that they are not based on transactions, although they may use transaction data.

Knowledge-based systems emphasize the separation of knowledge about the application from the processing of that knowledge. Processing typically uses different types of reasoning, such as analogical or qualitative reasoning, in addition to the classical logical type. A successful application of this type of system is to fault diagnosis, where, for example, large generators of electricity have their mechanical vibration patterns continuously analysed. The system is able to compare certain patterns of vibration against others and can issue a warning to avert damage or a breakdown.

SPECTRUM OF SYSTEMS

It may be useful to think of transaction processing systems and executive information systems as being at the extremes of a spectrum of information systems. This would allow an application such as a personnel application, where there is neither a high transaction frequency nor a great deal of very sophisticated processing, to be located somewhere in the middle.

ORGANIZATIONS AND APPLICATIONS — HISTORY AND TRENDS

The first users of computers, in the 1940s in the United States, Germany and the United Kingdom, were government owned or sponsored organizations, involved mainly in mathematical processing with a small amount of input information, solving equations for missile trajectories and decrypting secret messages. Other 'scientific' applications such as weather forecasting, by using a model of the earth's weather system, encouraged the development of FORTRAN in 1956 as the first high-level programming language.

ment, and will also define methods for applying the standards. Another task might be to conduct tests on software and ensure compliance with relevant legislation, such as the Data Protection or Computer Misuse Acts.

Technical services

This area contains specialist functions for the infrastructure of the computer department. Capacity planning is concerned with planning future hardware requirements, monitoring current hardware performance and carrying out benchmarking and acceptance tests for potential systems. The communications area is concerned with the technology of communication between, for example, user terminals and central computers, over the public switched telephone network or via private packet-switching networks. An alternative is using satellite transmission. Another area concerns short distance connections between devices, typically within a building, using local area networks (LANs).

The systems programming area has several responsibilities, namely hardware and software commissioning and monitoring the technical performance of computer systems. Systems programmers do not write programs that will form part of an information system, but instead install and tune the systems software supplied with the computer by its manufacturer or other packaged software used by many information systems.

The database area contains two specialists who affect other areas. The database administrator is responsible for data design and standards for databases shared by many organization users, while the data analyst performs a systems development task concerned with representing objects in user applications by data. This area has emerged only recently and is interesting as it is a specialist task that impinges directly on some activities in systems development.

Customer services

Two activities may be mentioned here, training and technical authoring. Training involves the development and presentation of courses to users on computer systems that they will use. This might involve educating and supporting users in connection with packaged personal computer (PC) software they have bought. Technical authoring concerns the design, development and maintenance of manuals describing how users should operate systems, how they should report faults or suspected errors, procedures to follow for back-up of data and similar topics.

ORGANIZATION OF PRODUCERS

Traditionally, for commercial information systems, the producers were organized into a DP (data processing) department within the organization, often coming under the accounting function. This was due to the fact that the earliest applications were primarily financial. All of the development work was undertaken by this department, which also possessed all of the computing resources.

However, this situation has been changing, due to the price/performance ratio of PCs and the availability of standard software, and systems development is sometimes decentralized to users in other departments in the organization who wish to develop their applications themselves. Many organizations have set up what is known as an information centre to implement this concept, and hope thereby to develop applications that are relevant to their needs and to develop them more quickly than the DP department. Systems that will be shared by many departments are still developed in the computer department, and computer staff are expected to develop standards for, or to coordinate the work of, the users in the different departments. It is still too early to say whether this form of organization has been successful.

Some large organizations, which may have many regional offices, factories or depots, often decentralize or distribute various functions of the computer department to such local facilities.

Where an organization produces its own information systems, the systems are said to have been developed in-house. The larger organizations still maintain large numbers of computer staff for this purpose. However, smaller organizations often approach software houses or management consultancies to develop information systems for them to their specification. Such software is termed bespoke software. Alternatively, organizations may buy a licence to use pre-written software, termed software packages, if the computing part of their desired system is fairly standard.

Information

It will not have escaped the reader that we are already in Chapter 2 and we have not yet defined what we mean by the term information, although we have defined information system, so this section will briefly discuss some problems with a definition of information.

DATA AND INFORMATION

Introduction

We could express the relationship between data and information in the following way:

$$Data + meaning = information$$

What we mean by this is that an item of data only becomes information when it is given a meaning. For example, the data 'UMIST' becomes information concerning the University of Manchester Institute of Science and Technology if an individual gives this meaning to the data.

How certain can we be that individuals give the same meaning to the same items of data? We shall briefly look at two contrasting paradigms of information (Harrington, 1991) in the organization that explores this question.

Two paradigms of information

- *Resource-driven paradigm*. This assumes that information is a vital organizational resource, to be managed with other resources such as capital and machinery. Information is basically unchanging, it is objective, it exists independently of its receiver and it does not change during transmission. The value of information is constant throughout the organization.
- *Perception-driven paradigm*. In contrast, this assumes that information does not exist beyond the perceptions of the receiver. It emphasizes that data is interpreted by individuals to become information and that interpretations will thus vary, on account of the different psychological and social factors inherent in individuals. The meaning of a data item also extends to its value, and perceptions of the value of information will also vary. The implication of this is that information systems cannot exist, as information is transient and subjectivity determines information. An information system is only a data system.

Formal and informal information systems

Although these two paradigms seem at opposite poles, they are reconciled in actual organizations as individuals are trained to interpret data in information systems consistently; that is an item of transmitted information will be interpreted consistently by different receivers. Information in information systems is often termed *formal*, or structured, information.

However, it is not clear to what extent such training may be successful. In work on management activities, for example, Mintzberg found that although managers were constantly receiving and transmitting information, the great proportion of this was *informal* information, as opposed to formal information (Mintzberg, 1973). One conclusion that may be drawn is that formal information, which is 'public' in that there is organizational agreement on its interpretation, may be of only limited use to managers.

VALUE OF INFORMATION

Relative value

One consequence of our awareness of the perception-driven paradigm is to realize that the *value* of information is perhaps as subjective as its meaning.

This is clearly important if we are attempting to build an information system by determining information needs, and we find that perceptions vary widely as to the value of different types of information.

A problem also to be faced is that the value of information is never absolute, but always relative to changing circumstances, such as the availability of other types of information or the situation the organization is in (such as profitable or not profitable). For example, what value should be placed on information concerning the economic state of MacAdam's competitors? Is this more important than information concerning the types of customers they have? Which information system should we build first?

Normative and realistic values

The normative value of information is based on decision theory, and measures the value of information based on the benefits that will be realized by possession of the information. Probability theory is used as all benefits have to be estimated. In practice, however, except for very structured situations, a normative value cannot be obtained as the probabilities cannot be estimated with any degree of accuracy.

A realistic value of information can, in theory, be obtained by estimating the difference in organizational performance based on the absence or presence of the information. Although one of the attractions of this approach is that it solves the problem of measuring information value by using economic indicators of organizational performance, it is very difficult in practice to set up an experimental situation in an organization that keeps all factors constant except those related to the information.

Reduction of uncertainty

Another approach to information value holds that the key factor in deciding what information is useful in an organization is the extent to which the information reduces uncertainty about an aspect of the organization. For example, Lucas states that 'information is some tangible or intangible entity that reduces uncertainty about a state or event' (Lucas, 1985). A problem with this idea is that there is no way of telling how detailed the information should be, as uncertainty is a continuous variable. For example, should a stock control manager know about all stock levels every week, or every hour?

SUMMARY

The practical conclusions of this discussion on information are that it may not be quite so easy to obtain agreement between users over the meanings to attribute to data. The analyst should in fact expect that different individuals will interpret data differently, although there is less scope for such variations in interpretation with functional information, where employees may be more or more of these systems, using one of the often large number of personal computer systems attached to the basic systems.

Summary

This chapter has discussed information systems from three perspectives. Different types of systems with their related applications were briefly described. The contents of an information system were then discussed, under the headings of people and procedures, and information. Some information system case studies and then examples of typical information systems brought together many points made earlier.

Discussion questions

1. Describe the notion of a transaction and give examples of a transaction in an insurance office, in an amusement park, in a garage and in a building society.
2. Discuss the type of information recorded and made available by a transaction processing system in each of the organizations mentioned in Question 1 above.
3. Identify the distinction between transaction processing systems and decision support systems.
4. What is the distinction between data and information? What are the problems and significance in determining the value of information?

References

Curtis, W., H. Krasner and N. Iscoe (1988) 'A field study of the software design process for large systems', *Communications of the ACM*, vol. 31, no. 11, pp. 1268–1287.
Davis, G. B. and M. H. Olson (1985) *Management Information Systems: Conceptual Foundations, Structure, Development*, 2nd edn, McGraw-Hill, New York.
Friedman, A. L. and D. S. Cornford (1989) *Computer Systems Development: History, Organization and Implementation*, Wiley, Chichester.
Harrington, J. (1991) *Organizational Structure and Information Technology*, Prentice-Hall, London.
Lucas, Jr, H. C. (1985) *The Analysis, Design and Implementation of Information Systems*, McGraw-Hill, New York.
Mintzberg, H. (1973) *The Nature of Managerial Work*, Harper and Row, New York.
Simon, H. A. (1960) *The New Science of Management Decision*, Harper and Brothers, New York.

OUTPUT CHECKING

The control loop shown in the diagram senses system output using the sensor component. The control device then compares some characteristic of the output to a desired standard. In its simplest form, any difference causes a corrective input to be generated by the activating unit to the functional system process so that the output will be nearer the standard. However, control systems may also alter processes or, for self-organizing systems, may alter system aims.

A common example is where the temperature of a heated room is checked against a standard temperature and any difference initiates a process to bring the room nearer to the standard. A central heating thermostat constitutes a control system that senses the temperature of the air warmed by the boiler and radiators, compares that temperature to a desired standard (usually manually set in the thermostat) and sends an input to the system to start heating, if the temperature falls below the desired standard.

An example where processes are altered is in the post room, where the number of items output per hour is monitored. If it is less than some standard figure the control device takes action and extra staff can be assigned to processes.

This type of control system is termed a negative feedback system, as it samples the output of the functional system and seeks to reduce fluctuations around a standard. A positive feedback system does the opposite — if the output characteristic deviates from the standard then the system processes are repeated or increased so that the deviation increases. We will be concerned only with negative feedback control systems.

INPUT CHECKING

Control systems frequently check the validity of system input as well as system output. This may be termed filtering. For example, in the post room, a check will be made that each input item has an address. If it has none, it cannot be processed and may be rejected.

DIFFICULTIES OF CONTROL

A control system may be seen as one that reduces the amount of uncertainty in the system. However, controlling complex systems may be a problem, as the law of requisite variety, in a simplified form, says that to control each possible state of the system elements, there must be a corresponding control state. In addition, it is necessary to receive and transmit control information to and from system elements, imposing a substantial information processing task. To predict all system states may not be possible, especially if the system under consideration is relatively open, with unpredictable inputs.

A solution often adopted is to use a human computer system for the control system, where the computer can generate control responses for expected cases and where a human decision maker generates responses for unexpected situations.

Design system

As many systems are continuously under review and are being redesigned, then to give a wider picture we may include a design sub-system, shown in Fig. 3.4, as both functional and control sub-systems interact with it. The control system detects discrepancies between functional system output and objectives, and will produce input to the design system to change the functional system's processes, inputs/outputs or objectives. The design system will be producing outputs that are designs for a new process or control system.

State space approach

The *state space approach* is an additional concept for describing system behaviour, and considers objects and processes. Under this approach, at any given instant a system is considered to exist in a certain state. This state is defined or characterized by the relevant objects that exist in the system at that instant. For example, in the post room, the state at a given instant will be determined by objects such as the letters, packages, numbers and types of staff, and so on. An object need not only be a physical object, but may be, for example, a person's age.

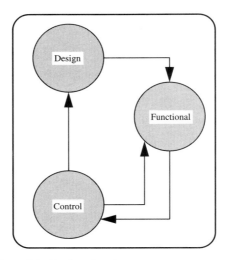

Figure 3.4   Design, functional and control systems

Such an approach is only applicable to *discrete* systems, that is systems that exist in separately identifiable states, and information systems are discrete systems. For example, in a system that models deposits and withdrawals from bank deposit accounts, a separately identifiable state occurs each time a deposit or withdrawal is made. In contrast, many other types of system are *continuous* systems, which are continually changing, often extremely rapidly. For such systems, it may not be possible to determine their states before the state has changed, for example the level of water in a reservoir.

An important type of process termed an *update process*, is viewed by this approach as a transformation of the system from one state to another. An update process is a process that makes changes to the system, for example adding a staff member to the post room, as opposed to a process that reports on the system state. For example, in a football match, an initial state could consist of 22 players and a score of 0–0. A subsequent state (achieved by the update process of goal scoring) could consist of the same number of players but a score of 1–0.

The state space approach is a useful extension to the basic systems approach which we shall use in our analysis of information systems.

Classification of systems

Many different attempts have been made to classify systems. For example, general systems theory considers open and closed systems, deterministic and non-deterministic systems, discrete and continuous systems, and abstract and physical systems, as we have seen.

The Open University definition (Open University, 1980) also discusses three types of systems, evolved, designed and model systems. Evolved systems have objectives that emerge from the evolutionary process. Although biological systems are prime examples, some social systems might fall into this category. Designed systems are designed by people specifically to achieve certain objectives, for example a telephone network.

Finally, model systems are systems embedded in 'messes' (complex situations that are poorly understood) that we can identify as an aid to structuring our thinking about the mess. Model systems are similar to the type of abstract system we identified at the beginning of this chapter. The first two types are termed recognized systems and the third type unrecognized or invisible systems.

The last classification we shall consider is taken from Checkland (1981). Checkland proposes four types of system:

1. *Natural systems*. For example, a human being.
2. *Designed physical systems*. For example, a washing machine.
3. *Designed abstract systems*. For example, philosophical or scientific systems such as utilitarianism or Darwinism.

4. *Human activity systems*. For example, a post room.

An information system under this classification would be a human activity system, while it would be a designed system under the Open University classification above. Existing classifications are thus based on different viewpoints, and no one classification is regarded as superior.

The systems approach — discussion

TYPES OF MODEL

The systems approach provides a *model*, based on systems concepts, which is claimed to be applicable to an unrestricted set of situations. We may define a model as *an abstract representation of reality*. The model is usually on a level of abstraction that is of interest, and there are three general types of model:

1. *Predictive* models are used for predicting the future and are therefore useful for planning. Given input concerning, for example, an event, they will determine the consequences of that event. A predictive model is also termed an analytic model, and many scientific models are predictive. For example, Newton's model of the laws of motion predicts that, if you drop an object from rest, its velocity at any time t will be ft, where f is the acceleration due to gravity. In a computing context, an analytic model of the performance of a certain type of file might be that the rate of access A to the file varied inversely with the square of the number of records N, so that $A = 1/N^2$.
2. *Normative* or optimizing models suggest the best action to be taken in a given situation. They contain norms or suggestions as to how systems ought to behave. Linear programming problems may be solved with normative models.
3. The systems approach provides the open systems model, which is the last type of model, the *descriptive* model. As its name implies, such a model allows for a description which is useful for leading to deeper understanding of a situation under investigation. This model is based on the classic 'divide and conquer' principle, where a situation is divided into smaller parts using the concepts of input/output and process and partitioned from other systems using the concepts of systems boundary and environment. The model further provides a typology of systems, from open to closed, depending on the degree of interaction with the environment.

It is important to realize that the model underlying the systems approach does not predict anything; that is it cannot predict that certain system behaviour will occur for a given input, for example. It is merely an approach for a description of situations in systems terms for clearer understanding. We shall use descriptive models almost exclusively for describing and specifying information systems.

PROBLEMS

The first problem with the systems approach is that, although individual understanding may be improved by using systems concepts, the analysis of a given situation by different individuals may result in quite different systems. This is a consequence both of the extreme level of generality of the underlying model and the fact that, in situations involving humans, it is difficult to agree on the facts of the situation. It is thus obviously a problem if the situation cannot be described precisely and objectively.

Contrast this 'soft' situation with a 'hard' situation, for example, the forces acting on a train going uphill. Here the situation may be described objectively by a predictive model incorporating the laws of motion. We are not so fortunate with the systems approach, as the systems we wish to describe are not governed by such laws.

The main problem with multiple views of a problem situation is: which one is the correct view? Is there more than one correct view? The resolution of multiple views into one common, acceptable-to-all view is often a difficult process. A particular aspect of this problem is that a system may be decomposed in different ways into sub-systems. These may be so different that it would appear that different systems are being described. This is often due to different individuals decomposing a system to different levels of abstraction. An associated problem is that the ease of obtaining widely differing descriptions does not lead to confidence in the approach.

A second problem is that a process is always considered to involve transformation of inputs into outputs. However, inputs that control or affect the process are often required and are not transformed. The approach might refute this criticism on the grounds that the approach is sufficiently general to include these different types of input or processes.

A third problem is that processes and objects may be shared by more than one system and the system boundary is often arbitrarily defined, so it may be difficult to know whether to include a process or object. For example, should machinery used in the post room be included in the post room system if it is also used by others for training purposes? Should the delivery process whereby letters are input to the post room be included? Sometimes there is a problem about including what may be termed general knowledge. For example, should the location of the post room be kept in the system description, as everyone knows it?

The fourth problem is that no fixed method is available for applying the concepts to the real world. Should we determine inputs and outputs first, or processes? How do we set the boundary? A different system might result depending on the way in which we proceed.

DISADVANTAGES OF SYSTEMS APPROACH

1. Its application to the real world results in a non-unique description; that is different applications of the approach give different results. There are problems with objective descriptions of processes and boundaries.
2. The model may be incomplete or inaccurate when applied, as there is no guidance to the level of detail required for a description.
3. No method is suggested to apply the approach.

ADVANTAGES OF SYSTEMS APPROACH

1. The approach provides an informal start to understand and describe a situation, using intuitively familiar notions of input, output and process.
2. The decomposition of processes is a useful method of analysing complex processes.
3. The notion of the importance of a control system often highlights the fact that one is missing or ineffective.
4. A system description is often simple enough for it to be used as a tool for communication between individuals, accompanied by a suitable explanation.

Summary

The systems approach described in this chapter is based upon the following concepts and assumptions.

Firstly, basic systems concepts are that:

1. A system may be described in terms of these components: input, output, process, boundary and environment.
2. Input and output are regarded as consisting of static objects, transformed by dynamic processes. Examples of objects are letters, people and so on.

Secondly, we note the need for a control and a functional system:

3. A control system, for checking input, comparing output and adjusting functional system behaviour, if required, is essential for controlling unpredictable systems. Information systems are a major type of unpredictable system.

Thirdly, we add the state space approach:

4. The state space approach views processes (update processes) as transformations between one state and another. A state at a given instant consists of the objects in the system at that instant.

The systems approach may be criticized, chiefly for its subjectivity and its high level of generality, but it possesses certain advantages, such as providing an informal start to understanding and analysis.

Discussion questions

1. Describe briefly, using an example, the components of an open system.
2. Give the Open University definition of a system. How would you criticize the Open University definition?
3. Are the examples below systems according to the Open University definition? Explain your answer.
    (a) A briefcase
    (b) Set theory
    (c) A firework
4. How would you classify the following systems, using the classification due to the Open University?
    (a) A map of England
    (b) A filing cabinet
    (c) A water molecule
    (d) An end-of-term party
5. Describe the main elements of the state space approach. Why is this important for analysing information systems?
6. Give some examples of information systems that are relatively closed and some that are relatively open. Do you agree that these are characterized by their inherent degree of uncertainty?
7. Explain why it is necessary to control information systems. What problems are there in trying to do this?
8. How would the following examples be classified, using Checkland's four-point systems classification?
    (a) PASCAL programming language definition
    (b) A television set
    (c) A hydrogen atom

References

Ackoff, R. L. (1971) 'Towards a system of system concepts' *Management Science*, vol. 17, no. 11, pp. 661–671.

Checkland, P. (1981) *Systems Thinking, Systems Practice*, Wiley, Chichester.

Checkland, P. and J. Scholes (1990) *Soft Systems Methodology in Action*, Wiley, Chichester.

Churchman, C. W. (1968) *The Systems Approach*, Dell Publishing Co., New York.

Open University (1980) *Systems Organization: The Management of Complexity*, Block 1, The Open University Press, Milton Keynes.

Von Bertalanffy L. (1968) *General System Theory: Foundations, Development, Applications*, Penguin, London.

4
Analysing organizations and information systems

Applying the systems approach

FUNCTIONAL SYSTEM AND CONTROL SYSTEMS

In Chapter 1, we used the terms application, information system and organization-related activities to gain an intuitive feel for information systems in organizations. However, these terms are widely used and are capable of many interpretations. In this chapter, we want to present a more precise description of information systems, so we will use instead the terms *functional system* and *control system*, with the general meaning given to them in the chapter on the systems approach. This is shown in Fig. 4.1.

The functional system is roughly equivalent to the application, and the control system contains the *information system* and the *management system*, consisting of management activities. We will show later how the information processing type of functional activities discussed in Chapter 1, concerned with information input and output, are usually considered as part of the information system. The view we are taking here emphasizes the use of information in the information system to assist management activities.

ORGANIZATIONAL MODELLING — THE OMNIS MODEL

We shall apply the systems approach to organizations to develop a model, based on Fig. 4.1, for describing information systems in a social and organizational context. The model is termed OMNIS — organizational modelling of information systems.

The model is a descriptive model, and allows us to analyse and describe an information system in an organization in more detail than the basic open systems model. We will use the model later, in the systems development

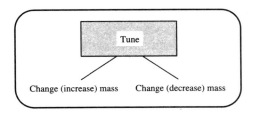

Figure 4.2 Refinement of the tune process

Support processes

Support processes are primitive processes, but they are grouped separately in the application. These are termed support processes as they are only indirectly concerned with actual production activities, but they operate on the objects and properties in the production system. For example, new customers are acquired, suppliers are discontinued, properties of products may be changed and so on. We show examples of such processes in Table 4.3.

When Eurobells Δ acquires a new customer or supplier, this may be regarded as a create process, adding the new object to the system. From time to time, properties of customers and suppliers, such as addresses and telephone numbers, will require a change process, and the organization may acquire a new supplier of a particular type of raw bronze (perhaps at a cheaper price). In this way the supplier property of raw material will be changed.

Table 4.3 Support processes affecting system objects and properties

Component	Type	Process
Customer	Object	Create
Supplier	Object	Delete
Bell pitch	Property	Change
Supplier address	Property	Change
Raw material supplier	Property	Change

EVENT

An *event* may be defined (Rock-Evans, 1989) as something that happens to an object in the organization. It is an occurrence of an activity that initiates a system process but is external to the system under consideration, for example, the arrival of a consignment of raw bronze or a customer placing an order.

In the functional system, processes are not continuously active, and the importance of events is that they are the triggers for groups of related processes. For example, when a customer rings with an order, this event triggers the package products process, and when a bell is out of tune, the tune process is triggered. As events and processes are closely related to each other in the system, we consider them jointly as constituting system *behaviour*.

RULE

Description

We have seen that an important part of describing a process is to define its inputs and outputs. However, in addition, processes must obey *rules*. For example, there might be a rule that the cast bell process is not allowed to produce a bell with a mass greater than 1500 kg. This rule governs an output, and an example of a rule governing input might be where the tune process is not allowed to take place on a bell (for quality control reasons) if the pitch of the bell is out of true by more than 1 per cent.

We express rules only in terms of object and property, and the type of expression we use is, for example: mass of bell $\leq$ 1500 kg. Many rules restrict property values or the numbers of instances of objects.

The main advantage to separating rules from processes is that we simplify both the description of processes and rules. When a process refers to an object or property, it must not violate any rule associated with the object or property.

Definition

> A rule is defined as a restriction on the system states (expressed in terms of objects and properties) that may exist.

Rules (together with the definition of objects and properties) thus define allowed system states. It is easier to define the allowable states by general assertions in the form of rules than by enumerating all the object and property instances that may exist.

The above is a static view of a rule, and a dynamic view is that it is a restriction on the inputs or outputs of a process. This view is obviously implied by the static view, as processes obeying the rules will have only allowed system states as input and will produce only allowed system states as output.

It can sometimes be difficult to identify rules, as they may only be implicit, being embedded in the detail of the way in which processes are actually carried out. For example, the bell moulds in Eurobells △ may be such that they can only hold so much mixed bronze, so the cast bell process is physically prevented from ever violating the rule that the mass of bell $\leq$ 1500 kg.

SUMMARY

Figure 4.3 summarizes our analysis of the functional system.

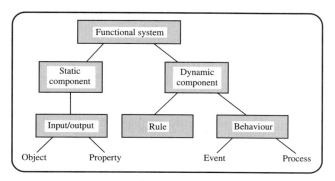

Figure 4.3 Functional system components

Control system

NEED FOR A CONTROL SYSTEM

For a large and complex social organization, no theory of organization exists that will allow us to predict its behaviour. When we discussed the systems approach in Chapter 3, we saw that such unpredictable systems need a control system to ensure that they function correctly.

CONTROL WITH AN INFORMATION SYSTEM

In the simple example of a control system given in Chapter 3, a thermostat controlled the temperature of a room. An important aspect of its operation was that it sensed a part of the functional system (the output room temperature) directly. However, in organizations, it is generally more efficient to have available relevant information about the functional system in an information system than it is to check the functional system directly. The reason for this may be illustrated by an example from Eurobells △ concerning its control over the amount of stock (raw material, products) it keeps in its stores. This is affected by processes, such as receive raw bronze and ship packaged product, from Table 4.1.

If the stock manager wants to know how many of these items are in stock at the end of every day, he or she can walk round the stores and count each item directly. However, this is a tedious, inefficient and error-prone job if the stock quantity is large.

A more efficient way, using an information system, is to start with an initial count of items, kept perhaps on card index or computer file, and to have procedures that record the stock items in and out of the store. These figures may then be added to or subtracted from the initial count, giving an up-to-date total of the amount of stock. The key to this approach is that the information in the information system is checked, rather than the items in the store. We assume that procedures have been carefully designed so that the information accurately represents the relevant part of the functional system.

CONTROL REQUIRES INFORMATION AND MANAGEMENT

Naturally, it is not enough to have only an information system. For control, the information in the information system is used to monitor the behaviour of the functional system, and changes are then proposed and made, if required. Such management activities take place in what we term a *management system*, shown in outline in Fig. 4.1. In the example above, a manager may use the stock control information in the information system to order an increase in production of treble bells, if their quantity has fallen below a certain point.

This example has concerned checking the state of the organization against a standard, and taking appropriate action. We shall see in the next chapter that other types of management activities exist which also use information from the information system; for example hiring and training staff or deciding on new products. As such activities are intended to have a positive effect on the organization, we may see them as control activities, in the broad sense of the word.

We therefore define the control system as consisting of an *information* and a *management system*, as shown in Fig. 4.4. The management system will be discussed in more detail in the next chapter.

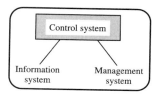

Figure 4.4 Components of the control system

Table 4.4 Data records representing objects and properties of two customers — the London Symphony Orchestra and Campanile Priory — and of the supplier Bronze Age

Object name	Properties (address, credit limit, . . .)
LSO	Barbican, London, 20 000, . . .
Campanile Priory	Paean Park, Berkshire, 1000, . . .

Object name	Properties (address, telephone number, . . .)
Bronze Age	Bronze Buildings, Brasilia, 09 23 45 63, . . .

An instance of an object is usually represented by a unique name, and the values of properties for a given object are usually stored together in a record, along with the object name. Sometimes a code is used instead of a name. A collection of these records is termed a file, which we may visualize as a particular area of the information base. Table 4.4 shows how some of the objects and properties from Eurobells △ are represented by data.

Processes

To represent the functional system processes, simulation system processes create, change and delete the objects and properties in the information base. The processes may be performed by humans or by computers, depending on the type of information system being used. Table 4.5 shows some processes together with the effect on the relevant objects.

MESSAGE SYSTEM

This system is concerned with retrieving information, in the form of messages, from the information base. It reports on the state of the simulation system and contains processes that provide the user with the required

Table 4.5 Representing processes in the simulation system

Functional system	Simulation system
Tune bell	Change the mass property of bell
Receive raw bronze	Add to quantity of raw material
Create a new customer	Create process creates a customer record and inserts values for properties
Delete product	The record representing that product is deleted
Delete supplier of raw material	Delete supplier property value on raw material record

information, in terms of objects and properties. Information may also be derived, according to *derivation expressions*, if it is not held explicitly in the information base. For example, the average credit limit of customers may be obtained by adding all customer credit limits and dividing by the number of customers. Processes may be human processes, manually obtaining information and typing reports, or computer processes that search files and send information to computer screens.

Information needs

The type of information that can be obtained from the information base is, for example:

- *Objects*. How many customers does Eurobells △ have currently? How many treble bells do we have in stock or on order? Is Bronzecheek a current supplier?
- *Properties*. What is the order date of Clapper CC13? How many products will be finished this year? What is the address of Crazy Casting? Is the credit limit of Oedipus' orchestra £300? What supplier(s) supply raw material type F56/P7? Have we ever supplied Boinnng! Monastery with a product?

HUMAN COMPUTER SYSTEM

Input system

This system keeps the simulation system in step with the functional system, and we can distinguish two types of processes within the input system, *data capture* and *transaction input*:

- *Data capture*. The occurrence of an event in the functional system initiates a data capture process. For example, when the receive raw bronze event occurs at Eurobells△, the event causes a data capture process to start, recording relevant details about the event, such as the supplier and the quantity.
- *Transaction input*. The next stage is to initiate the process in the simulation system that represents the corresponding functional system process. This is done by a transaction input process, which uses the data 'captured' by the data capture process to provide input to the simulation system process.
- *Coupling between functional and simulation systems*. Coupling refers to the degree of time synchronization that exists between these two systems. A simulation system process need not occur instantaneously (or as soon afterwards as possible) with its corresponding event. The user require-

ments will determine how up to date the simulation system is to be. Thus, the data capture and transaction input processes could, at one extreme, be merged. At another extreme, they might be separated by a time factor of hours, days or even weeks.

Output system

In a computer-based information system, message system processes will be computer processes. In order to interact with these processes, procedures for humans as well as computers will exist. In addition, message formats will be required, for example, screen formats for input and output, menus, report layouts and so on. We may distinguish broadly between two types of output system process, the *query* and the *report*.

A query is an inquiry for a small amount of information that may be made at any time, and expects a result after a short interval, for example within a minute. The query is generated by human intervention. A report is usually a planned request for a larger amount of information that takes place periodically, such as at the end of the day or week. Human intervention may not be necessary and a report may be produced automatically from a computer-based information system. For example, a query will find out a supplier address, given a supplier name, while a monthly printed list of all current suppliers and their properties would be produced by a report.

OMNIS model

Figure 4.7 integrates the topics that we have covered in this chapter. It shows the OMNIS model, which we shall use both to analyse organizations and as a base to design and build computer-based information systems that operate within an organizational context.

OMNIS AIMS

The basic aim of OMNIS is to analyse a description of an organization so that relevant facts are separated out using the 'pigeon-holes' or categories of the model, such as functional system, simulation system, information base and so on. The structured description that results is a simpler description than the original as irrelevant material has been ignored. The description may be used to detect incompleteness or inconsistencies, and may also be used as the basis for checking, with the users, that the analysis is correct. We deliberately avoid the use of a precise language for the analysis.

To apply OMNIS properly, you should understand the following principles and method.

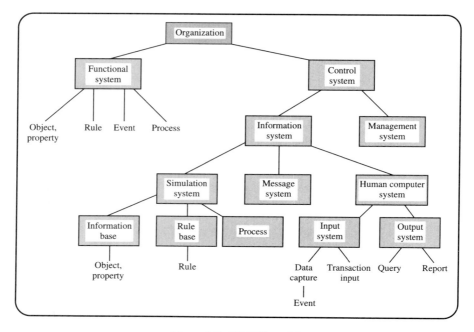

Figure 4.7   OMNIS model

OMNIS PRINCIPLES

1. *Non-redundancy*. Model a fact only once. Do not duplicate facts.
2. *Completeness*. Model all the relevant facts.
3. *Accuracy*. This principle is the corollary of the preceding principle. Do not 'invent' facts. Model only what is in a description.
4. *Clarity*. Model a fact as an object, rather than a property, as an object is easier to see in the result. Only model important properties, as others will be added later.

OMNIS METHOD

1. Read the description of the organization. Broadly identify the scope of the functional and control systems.
2. If message system (MES) processes are fairly obvious, then proceed as follows:
   2.1 List and number MES processes.
   2.2 Identify information base (IB) elements referenced by the MES processes and classify as objects and their properties.
   2.3 Identify simulation system (SIS) processes that update the IB elements:

      2.3.1   List and number processes.

      2.3.2   Refine processes where necessary.

      2.3.3   Identify support processes.

  2.4   Identify human computer system (HCS) processes:

      2.4.1   Input system (IS) processes: list and number events and associated SIS processes.

      2.4.2   Output system (OS) processes: refer to MES processes initiated, and specify whether they are query or report.

3.   If IB elements are fairly obvious then proceed as follows:

  3.1   Classify IB elements as objects and their properties.

  3.2   Identify, list and number MES processes that reference the IB elements.

  3.3   As 2.3 above.

  3.4   As 2.4 above.

4.   Verification. Check your result as follows:

  4.1   SIS and MES processes must not reference elements that are not in the IB.

  4.2   If any IB element is not referenced, check the description again.

  4.3   All MES processes should be referenced by HCS(OS) processes.

  4.4   All SIS processes should be referenced by HCS(IS) events and processes.

Summary

1.   In the functional system, the static component is refined into object and property. The states of the functional system are defined in these terms.

2.   In the functional system, processes are update processes, which change the system state. All processes can be refined into primitive processes, which consist of create, change or delete processes, and others, which operate on the objects constituting the system state. Events trigger processes.

3.   The control system consists of two systems, the information system and the management system. Broadly, the control system controls the functional system, using the information and management systems.

4.   The management system consists of management activities, performed by users. These activities require information in the information system.

5.   The information system consists of three systems, the simulation system, the message system and the human computer system.

6.   The simulation system maintains and stores information representing the states of the functional system and consists of processes representing the functional system processes, while the functional system states are represented by the objects stored in the information base.

7. The message system contains processes that only read the contents of the information base, transmitting messages concerning system states to the management system.
8. The human computer system consists of the input system and the output system.
9. The input system keeps the simulation system up to date with the functional system, with the concept of event. The time coupling between the two systems may not be instantaneous.
10. The output system allows users access to the message system.
11. Objects are represented in the information base with data, while processes are represented by human, computer or some other type of process.

Discussion questions

1. Explain why a model such as the OMNIS model is useful for describing information systems.
2. Without consulting the text, list the constituent sub-systems into which OMNIS analyses the organization.
3. Describe the relationship between the functional, information and management systems.
4. Describe the relationship between objects and properties, system state and update processes.
5. What is an update process and what are the three types of primitive process into which it may be refined?
6. Explain how rules are related to (a) objects and properties, (b) processes. Are rules used by message system processes?
7. Explain the concept of event and how it is used to synchronize the functional and information systems.
8. What type of process is a support process?
9. Describe the functions of the input and output systems within the human computer system.
10. What relationship does the information system have to the control system?

References

Mintzberg, H. (1979) *The Structure of Organizations*, Prentice-Hall, Englewood Cliffs, NJ.

Rock-Evans, R. (1989) *A Simple Introduction to Data and Activity Analysis*, Computer Weekly Press, London.

Case Study 1 — the fishing fleet 'La Perle'

INTRODUCTION

The case study has been designed as a worked example for demonstrating how OMNIS is applied to the analysis of information systems in organizations. The input to the analysis that we shall use is a descriptive organizational document written in English.

This case study concerns an organization that contains an information system. Such documents are commonly used when new systems are required, and are expressed from the point of view of a user (or potential user) of the system who works for the organization concerned. From the case study you will become familiar with (a) how organizations and information systems are described from a user viewpoint, (b) how to apply the model to the organization and (c) which format to use for presenting the results.

BACKGROUND

The fishing fleet 'La Perle' consists of a fleet of modern boats based on the port of St Quay-Portrieux, on the northern coast of Brittany in France. The fleet is interested in two main sea products, fish and seaweed. The fish is sold to humans while the seaweed is finding increasing favour as an organic fertilizer for the artichoke-growing area around Roscoff. The owner of the fleet, Capitaine Yeux d'Oiseau, is a young woman who has inherited the business from her father, and, as a recent engineering graduate in fish processing and informatics from the University of Brest, has developed an information system to make the business more efficient. She uses a new desktop PC.

The quantity of fish caught and sold is recorded on a daily basis and accumulated to a monthly total. Owing to the rapid decomposition of fish, no stocks are kept after the day of the catch. The quantity of seaweed caught and sold is recorded daily only. It is also important to record which customers buy fish and seaweed and in what quantities, again on a daily basis. A monthly total of fish sales to customers is provided. Customer details such as name and address are also included.

A large percentage of costs involves maintenance of the fleet. Each boat has a record kept of its name, last maintenance date and cost. The cost has a limit of FF 50 000. Information is kept concerning the docks where boats have been maintained. The names and addresses of the docks are also kept. In addition, information relating to repairs to nets is kept, including net number, date of purchase and number of repairs.

The way the system works is that each boat captain writes down the quantity of fish and seaweed caught every day and sends these details to the office staff in the harbour. Staff who sell the catch in the daily market note the quantities of the various items sold and to whom.

The type of information that the Capitaine has available includes: How many fish were caught and sold today and last month? What customers bought seaweed today? Who is my biggest customer this month for fish? When did we last maintain the boat *Pisces*? How many times have we repaired net number 14?

With the information, she is able to offer the best customers a discount if they guarantee to buy a certain quantity of the catch on a regular basis. Staff decide when boats should be maintained, based on yearly maintenance intervals, and when nets should be repaired. After 10 repairs, a net is discarded.

Using this description of the organization and the information system you are required to apply the OMNIS model and produce a description of the information system.

SOLUTION — INFORMATION SYSTEM

SIMULATION SYSTEM

Information base

Object	*Properties*
FISH	Quantity caught (daily)
	Quantity sold (daily)
	Quantity caught (monthly)
	Quantity sold (monthly)
SEAWEED	Quantity caught (daily)
	Quantity sold (daily)
DOCK	Name
BOAT	Name
	Last maintenance cost
	Last maintenance date
	Dock name
NET	Number
	Number of repairs
	Date of purchase
CUSTOMER	Name
	Quantity fish bought (daily)
	Quantity seaweed bought (daily)
	Quantity fish bought (monthly)

Rule base

Number of repairs of NET ≤ 10
Last maintenance cost of BOAT ≤ 50 000

Process	*Refinement*
1. Catch fish	Replace quantity caught (daily) property of FISH Replace quantity caught (monthly) property of FISH
2. Sell fish	Replace quantity sold (daily) property of FISH Replace quantity fish bought (daily) property of CUSTOMER Add to quantity sold (monthly) property of FISH Add to quantity fish bought (monthly) property of CUSTOMER
3. Catch seaweed	Replace quantity caught (daily) property of SEAWEED
4. Sell seaweed	Replace quantity sold (daily) property of SEAWEED Replace quantity seaweed bought (daily) property of CUSTOMER
5. Maintain boat	Replace last maintenance date property of BOAT Replace last maintenance cost property of BOAT Replace dock name property of BOAT
6. Repair net	Add to number of repairs property of NET
7. Set monthly totals	Set to zero monthly totals on FISH and CUSTOMER
Create, change, delete	Dock, customer, net, boat

MESSAGE SYSTEM

1. List FISH quantity sold and caught
2. List name of CUSTOMER with quantity seaweed bought (daily) > 0
3. List name of CUSTOMER with largest quantity fish bought (monthly)
4. List last maintenance date of BOAT with name 'BOATNAME'
5. List number of repairs of NET with number 'NETNUMBER'

HUMAN COMPUTER SYSTEM

Input system

Event	*Process*
1. After all catch is landed	Catch fish (1) and Catch seaweed (3)
2. At end of day	Sell fish (2) and Sell seaweed (4)
3. Boat returns from dock	Maintain boat (5)
4. When net is old	Repair net (6)
5. At month end	Set monthly totals (7)

Output system

1. Daily report process for MES processes 1 and 2
2. Query process for MES processes 3, 4 and 5

NOTES

1. The harbour, staff, Capitaine, artichokes and so on are not modelled in the information system, as we do not wish to record information about them. The Capitaine, staff and the boat captains are users — they perform activities in the HCS and the management system. However, this part of the description is indispensable for understanding the social and organizational context into which the information system will fit, and will be used to plan the management system.

   Management system activities might be, for example: (a) offer best customers a discount in return for guaranteed purchase, (b) send boats for maintenance at yearly intervals, (c) send nets for repair, (d) discard nets after 10 repairs.

   The desktop PC hardware and software is not modelled in the system, as it is only an implementation vehicle.
2. Objects FISH and SEAWEED are 'abstract' objects and have only one instance each. In contrast, BOAT or CUSTOMER are more typical and have many instances, representing the physical instances of boat and customer.
3. The model is incomplete, as no distinction is made between data capture and transaction input processes. Users should be queried about this.

5
Management system

Introduction

In the previous chapter, we formed a picture of the interaction between a functional system and a control system, which may be represented by the diagram in Fig. 5.1. This shows information from the information system being used in a management system to assist users with management activities. Many of these activities will result in changes to the functional system.

The production system example from Eurobells △ that we used gives a rather narrow view of the use of a control system, as it shows information being used only to monitor and control organizational behaviour. There are other types of management activity performed by users in the management system which we shall discuss below. This will give us an insight in to the kind of information that users require from the information system.

The management system is not as easily visible within an organization as the information system, which has a physical presence of its own, normally being located on computers, disks, printers, terminals and so on. This is because there may be many different individuals, in different departments or locations, who use the information for activities which themselves are not necessarily explicit.

Classical model of management activities

There are several models of management activities, but we shall examine only two contrasting ones. The classical model of management activities derives from the turn of the century French engineer and manager Henri Fayol (Fayol, 1949; Pugh, Hickson and Hinings, 1983), and is a normative model,

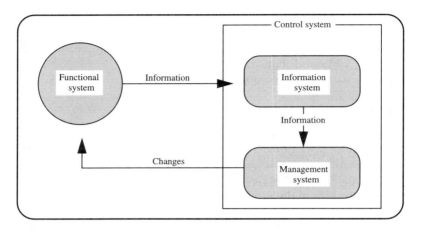

Figure 5.1   Functional and control systems showing control system components

that is, a model that sets out what it thinks managers ought to do. Fayol thought managers ought to:

1. *Plan*. Goal selection and policy definition for achieving goals for the future.
2. *Organize*. Creating organizational forms and structures from materials and humans.
3. *Direct* (originally, command). Leading and motivating employees and organizing their activities.
4. *Coordinate*. Ensuring that different organizational components are working together efficiently and harmoniously, for example synchronizing the work of two different departments, such as the production and packaging departments in Eurobells △.
5. *Control*. Ensuring that the organization behaves in accordance with established rules and expressed commands. In OMNIS terms, monitoring functional system behaviour, comparing the outputs with system objectives and adjusting the system if necessary.
6. *Staff*. Selecting and training employees.

The sixth activity, the staff activity, has been added more recently (Davis, 1974). These activities are general for all organizations.

Mintzberg model of managerial roles

Henry Mintzberg has produced a descriptive model of managerial roles, based on empirical research, where he 'shadowed' five senior executives in all aspects of their work (Mintzberg, 1973). In comparison to the classical model, the Mintzberg model is descriptive and is based on his empirical studies.

He found 10 roles, each role characterized by a set of similar managerial activities, and suggested that the roles be grouped into three areas, informational, interpersonal and decisional.

INFORMATIONAL

Managers are often key figures at the centre of networks of information; and there are three informational roles where the manager receives information as input and supplies it to others as output. In the *monitor* role, the manager obtains information, mostly current. In the *disseminator* role, information is transmitted to others in the organization. The information may concern areas outside as well as within the organization and may consist of fact or opinion. The third role is that of *spokesperson*, where the manager provides information concerning the organization to outsiders, such as the general public and those in influential positions.

INTERPERSONAL

Three interpersonal roles exist, all characterized by the involvement of the manager with people, internal as well as external to the organization. Firstly, managers may act as a *figurehead*, performing symbolic duties such as speaking at an annual dinner or a party given for a departing employee. The second role is that of a *leader*, concerned with staffing and motivation and training of individuals. Finally, the *liaison* role involves maintenance of contacts with individuals, for information acquisition purposes, both inside and outside the organization.

DECISIONAL

In the first of these roles, the manager is an *entrepreneur*, seeking for and initiating change in the organization to improve it in some way. Secondly, there is the *disturbance handler* role, where the manager responds to problems that arise. Thirdly, *resource allocation* involves deciding on the distribution of resources such as people, time and money to organizational tasks. Finally, disputes arise from time to time, perhaps between a manager's subordinates or between managers and subordinates; as a *negotiator*, the manager needs to resolve the disputes.

ROLES AND INFORMATION REQUIREMENTS

As we remarked in the introduction, we can see that management activities span a wider range than those of the production system example, which only focused on an activity that was a type of disturbance handling, using the

Mintzberg classification. In addition, it is clear that, in this model, information processing is at the heart of all management activities.

We may consider some examples of how the roles make use of information from an information system.

Informational

An example of an important dissemination activity is to send information, such as an annual balance sheet, to external individuals and bodies such as shareholders and investors. For example, Eurobells △ has to send a monthly profit and loss account to Eurobank$, who have loaned it £2 000 000 for recent expansion.

Interpersonal

A manager may, based on employment history, send employees on appropriate training courses if their history indicates that they need a certain skill. For example, Eurobells △ sends its more musical employees to different levels of campanology classes, so that the bell-ringing demonstrations used on marketing drives may maintain a high standard.

Decisional

Many examples may be seen here. As entrepreneur, a manager needs information on, for example, public response to a planned product. An early step in the marketing plans for a new size of bell at Eurobells △ is the generation, maintenance and analysis of survey data, from prospective and current customers, regarding their views on the planned bell. As a disturbance handler, managers may measure organization performance against objectives, as discussed earlier. As a resource allocator, a manager requires information about current organizational resources, to decide, for example, on funding for a project to produce a prototype of a new bell.

FEATURES OF THE MINTZBERG MODEL

One of the useful aspects of the model, for our study of information systems, is that the different types of role that are defined give us an intuitive understanding of the kind of information required to carry out the roles. We may note in passing, as a comment on the model, that the first informational role, where the manager monitors information, seems to overlap partially with the disturbance handler role.

The model emphasizes management activities, internal to an organization, that use the information for a purpose. Human computer system activities that merely retrieve the information from the information system to pass on to managers are not covered by the model.

Wide concept of management system

EXTERNAL USERS

We can extend the concept of management activities to external users also, as their actions may have an effect on the organization. This is shown in Fig. 5.2. External bodies such as Customs, Income Tax, shareholding institutions, banks, customers and suppliers may act as a disturbance handler to check some aspect of the organization against their own standard. Such a standard might be, for example, competitors' prices for similar products or the annual dividend for shareholders. In this way, the market-place may constitute an external control system.

Decisions and actions taken as a result of this checking process will often have an effect on the organization. An example of an indirect effect might be a slackening in demand for an organization's products when prices become too high, forcing organization management to consider a change. Direct effects may occur where there is more at stake. For example, the major partner in a loan consortium may demand changes to the board if a loss is declared in the annual accounts.

Not all information may be used in decisional roles. An example of an informational activity might be where information concerning the proportion of different ethnic groups among organizational employees is sent to an external body, for compilation and publication of a survey on organizational policy in the area of job discrimination.

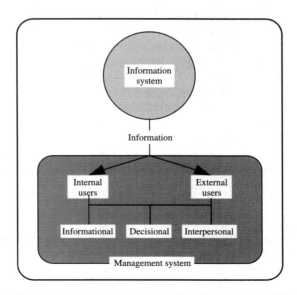

Figure 5.2 Management system components, showing external and internal users

The last point we may make is that not all individuals who fulfil a managerial role in fact have the job title manager, or even supervisor. There are many jobs entailing responsibility for close supervision of organizational affairs that may be performed by non-managerial employees.

DEFINITION OF MANAGEMENT SYSTEM

After our discussion, we can define a management system as follows:

> A management system consists of individuals or bodies, internal or external to the organization, who use information from the information system to assist in their management activities.

Relative importance of different management roles

Comparing the classical model with the Mintzberg model, activities 3 and 6 fit the interpersonal category and activities 1, 2, 4 and 5 are in the decisional category, while informational activities are not mentioned.

It is traditional that the information required and functions performed by management for decisional activities have received the most attention, as they have the greatest potential for bringing about improvements to organizations. However, a more modern view also emphasizes the importance of the informational activities.

Anthony's classification of decisions

Three types of decisions are made in organizations, according to Anthony (1965). These are:

1. *Strategic planning.* Strategic planning decisions are concerned with setting organizational objectives and deciding the type of business that the organization should be in. They are decisions with long-term implications which may require a substantial amount of organizational resources.
2. *Management control.* Management control decisions concern the acquisition of new resources, ensuring that they are used efficiently and effectively to accomplish organizational objectives.
3. *Operational control.* Operational control decisions use existing facilities and resources to check that specific organizational activities are carried out properly.

Although these decision types are not related to types of user by Anthony, it is generally assumed that strategic planning is carried out at the highest level of management, for example, by boards of directors in British organizations. Management control (often termed tactical management) is in the hands of middle managers, while overseers or supervisors form the lowest level of management, the operational control level.

Combining activities and levels

To illustrate the applicability of the discussions above, we may combine the Mintzberg model with Anthony's three levels of decision. For example:

1. *Operational control level*

Decisional (entrepreneur)	An overseer checks the quantity of certain raw materials and decides to reorder some that are known will be in demand soon.
Decisional (disturbance handler)	Production line A is critical and is falling behind schedule. The overseer transfers workers from a non-critical line to line A.
Interpersonal (leader)	An overseer assigns a new employee to an experienced workman to learn a specific job.

2. *Management control level*

Decisional (disturbance handler)	A manager analyses weekly production figures.
Decisional (resource allocation)	A manager analyses monthly costs of the profit centre by product and by production line. As a result, workers may be hired/fired or cheaper/more expensive raw materials may be ordered.
Informational (dissemination)	Every month, Eurobells △ has to send figures for industrial effluent dumped in the nearby river to the National Rivers Authority.

3. *Strategic planning level*

Decisional (entrepreneur)	Analyse product sales and initiate new products.

Management information requirements

INTRODUCTION

We may now attempt to systematize the different types of information required for management activities.

BASED ON THREE LEVELS OF DECISION

Information may be characterized in terms of its level of detail and its timeliness, on the basis that different levels of management activity require different levels of detail of information, concerning shorter or longer periods of time. According to this view, the lowest, operational control level of management requires very detailed information that is very up to date, for example, information concerning the system that is less than a day old; we may term this *operational information*.

Middle management would require information concerning a longer time scale, for example weekly or monthly, and would only require, for example, transaction levels to be summarized for that week or month. This can be termed *managerial information*. Finally, the strategic planning level would require information to be summarized at a higher level of abstraction, concerning a quarterly or half-yearly period, and we will term this *strategic information*.

For external information, reports such as yearly balance sheets for shareholders and institutional investors are a good example of strategic information, perhaps at its most condensed, and managerial information might be produced for the tax authorities, such as annual earnings and deductions for each employee.

FUNCTIONAL INFORMATION AS A TYPE OF MANAGEMENT INFORMATION

We should note that functional information, discussed in Chapter 1, may also have a management use, as it is required in this role by some external users. For example, information relating to a single transaction in the form of an invoice, remittance advice, despatch note or receipt may be input to external decisional activities. External users use a despatch note to check that goods delivered are the right goods or they can use a receipt to compare prices with other organizations.

This information is typically disseminated by clerical employees, and, as we have seen, such information processing activities are carried out in the human computer system. With increasing use of electronic data interchange (EDI), involving, for example, the transmission of invoice information directly to information systems belonging to customers, no individual may be involved at all.

Table 5.1   Four types of information required by the management system (adapted from Lucas, 1985)

Information characteristic	Information type			
	Strategic	Managerial	Operational	Functional
Source	External	Internal	Internal	Internal
Accuracy	Less important	Important	Very important	Vital
Level of detail	Very summarized	Summarized	Detailed (high volume)	Detailed (low volume)
Time interval covered	Long	Medium	Short	Very short
Timeliness	Not so recent	Fairly recent	Up to date	Up to date
Effect of resulting decision	Long term	Medium term	Short term	Short term

FOUR TYPES OF INFORMATION

To summarize, we identify four types of information, which are hierarchically related:

1. Strategic
2. Managerial
3. Operational
4. Functional

Table 5.1 shows characteristics of information that are required for each level of management decision, as suggested by Anthony. The importance of this type of table is the help it gives us, when designing information and related management systems, in choosing the right type of information for the right type of decision. A common error, for example, is for a designer to provide a senior manager, required to make control decisions, with information that is too detailed.

Summary

This chapter looked at two models of activities performed by managers in organizations, the classical and the Mintzberg models. The classical model consisted of six types of normative management activity, while the more recent Mintzberg model described 10 different types of management system activity in three role categories. We used this to produce a model of management system activities that differentiates between internal and external users, also defining a management system.

We then discussed the three levels of management decision, according to Anthony, and showed how these could be combined with the Mintzberg activities, giving a matrix of activities on different levels. On the basis of this discussion, we made some inferences about the types of information and their characteristics required for management activities, including functional information, and we presented a table showing the four different types of information, based on six information characteristics, required for different management system activities.

Discussion questions

1. Of the 10 managerial roles identified by Mintzberg, to what extent do you think they process informal, as opposed to formal, information?
2. Apart from transaction processing systems, what other sources of information are implied by Mintzberg's managerial roles?
3. Looking at Table 5.1, could you use this to determine whether a given item of information was strategic as opposed to managerial?

References

Anthony, R. N. (1965) *Planning and Control Systems: A Framework for Analysis*, Harvard University Press, Boston, MA.

Davis, G. B. (1974) *Management Information Systems: Conceptual Foundations, Structure, and Development*, McGraw-Hill, London.

Fayol, H. (1949) *General and Industrial Management* (Translated by C. Storrs), Pitman, London.

Lucas, Jr, H. C. (1985) *The Analysis, Design and Implementation of Information Systems*, McGraw-Hill, New York.

Mintzberg, H. (1973) *The Nature of Managerial Work*, Harper and Row, New York.

Pugh, D. S., D. J. Hickson and C. R. Hinings (1983) *Writers on Organizations*, 3rd edn, Penguin Books, Harmondsworth, Middlesex.

Part Three
Hard Approach to Information Systems

Part three describes methods and techniques that are representative of what has come to be known as the *hard* approach to systems development. Broadly, this means that the methods and techniques assume that a problem to be solved has a logical or mathematical basis and that a computer system, which has its functions specified very clearly and in great detail, is in most cases a suitable solution.

Chapter 6 describes the traditional variant of the hard approach, discusses problems and shows how alternative approaches attempt to address these problems, while Chapter 7 describes the requirements determination phase, and again discusses the problems found with traditional approaches. It then mentions several methods that provide partial problem solutions.

Chapters 8 and 9 describe the analysis phase in detail, illustrating the object-oriented approach to entities, rules and processes.

Chapter 10 describes four well-known methods, Structured systems analysis, Information Engineering, JSD and SSADM, all of which use some of the techniques described so far, and then discusses the solutions these methods provide to a case study of a university library.

Finally, in Chapter 11, software tools for method support are described, focusing on one particular tool, AUTO-MATE PLUS, which closely follows the SSADM method. Again, problems posed and addressed are described.

6
The systems development process

Introduction

SYSTEMS DEVELOPMENT PROCESS

To bring an information system into existence, many different activities need to be undertaken. For all but the smallest systems, a team process is involved, where many different individuals, users and computer specialists work together to define user requirements, produce a system specification, select and integrate computer hardware, write computer software, test the system, train users, operate the system and so on. This process is known generally as the *systems development process*, which we shall subsequently refer to as the *process*.

As organizations are always changing, for example, by responding to changes in the environment or by acquiring improved technology, it is usual for information systems to be in a state of frequent modification, requiring replanning, software rewriting, retesting and so on. Hence, another name for the process is the *systems life cycle*.

HARD AND SOFT APPROACHES

There are several different views concerning the nature of the process, and in this book we identify two broad approaches, commonly termed the *hard* and the *soft* approaches. This chapter will discuss variants of the hard approach, with the soft approach being discussed in a later chapter.

Traditional approach

The *traditional approach* to the development process is perhaps the best known, and a version of this may be seen in Fig. 6.1. The process, in its first phase, attempts to determine the scope and type of system the user wants.

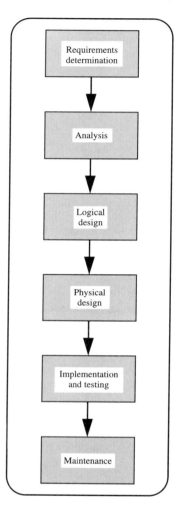

Figure 6.1 Traditional approach to systems development

This may involve a wide-ranging investigation into different kinds of application and system possibilities. The next phase analyses the system requirement into smaller parts, so that it can be understood and checked in detail, and then the system is designed and implemented.

It is useful to identify two important constituents, *phase tasks* and *phase products*, which we may define very generally as follows:

A phase task is an activity within a phase of the systems development process which produces a phase product.

A phase product is a description of a system on a certain level of detail.

The majority of approaches to the process consist of distinct activities separated into phases where each phase usually contains many stages, each producing one or more phase products. Most of the phase products are input to different stages in successive phases, and so the process may be seen as a transformation process, separated into phases, each of which adds further detail to an evolving description of the information system.

The terms used to describe the phases, as well as their content, are not yet fully standardized. Different authors will employ different names and draw the boundaries between the phases differently. This particularly applies to the first two phases.

Requirements determination

The *requirements determination* phase consists of several stages: *problem definition*, *feasibility study*, *requirements acquisition* and *requirements analysis*. The aim is to obtain a description of the user requirements which are expressed in terms of user concepts. In the problem definition and feasibility study stages, the outline of the desired system is defined. The requirements acquisition stage obtains user requirements in an unstructured form, while the requirements analysis stage uses a model such as the OMNIS model, introduced in Chapter 4, to analyse the requirements and to produce a phase product termed the *requirements specification*.

PROBLEM DEFINITION

The starting point for deciding to build a system is often the perception of a problem in an existing system which the user feels can be improved by the introduction of a computer-based information system. However, another reason might be to improve some aspect of the organization, for example, the speed of customer service, in the hope that this will put the organization ahead of the competition.

The aim of the activity is to define, on a high level of detail, the application for the desired information system and an indication of the advantages that will result. The problem definition is drawn up mainly by user management, although a computer specialist may be consulted, especially if the organization has a strategic IT plan into which new system proposals must fit.

FEASIBILITY STUDY

The feasibility study takes the problem definition and examines different alternatives that may exist for the desired system. For example, it may become apparent that computerization will bring no advantages and that a manual system will be more effective! More typically, there will be several alternatives for drawing the boundary between human and computer activities.

This activity is essential as it attempts to establish whether the proposed system will in fact bring about the advantages that are expected. It does this by anticipating costs and benefits of different kinds. An important aspect of this is to determine whether the cost of the system will be contained within the budget set by user management. Another important part is to set a boundary to the desired system. For example, a stock control system may be required for just one regional depot as a pilot study. Non-functional requirements, which are requirements concerned with, for example, resource constraints, data archiving and security, may be ascertained in this stage and developed in more detail later.

The feasibility study is usually carried out by systems analysts and results in a report containing different options for a proposed system with recommendations.

REQUIREMENTS ACQUISITION

Assuming that the feasibility study has identified and approved the type of system required, the next step is to acquire the system requirements in more detail. Requirements acquisition methods are used for obtaining the requirements about the desired system from users, and they are normally used by analysts or analyst/programmers. The four traditional methods are: (a) observation, (b) analysis of existing system documents, (c) analysis of desired system documentation, (d) interview or questionnaire.

The next chapter describes these methods and others that address some method problems, in more detail. Any of the methods may be used, separately or in combination, in this phase.

REQUIREMENTS ANALYSIS

It is becoming increasingly common to use a model to organize the facts that result from requirements acquisition into a more structured form. This makes it easier for analysts to check the requirements for properties such as completeness and inconsistency and to refer back queries to the users. The OMNIS model addresses this problem area, as it is on a user level, and we may use it in this stage.

PHASE TASKS AND PRODUCTS

We will term the phase product from the requirements acquisition stage the *statement of requirements*. This is a description, on a user level of abstraction, of the desired system, and is typically in an unstructured, natural language form.

The requirements analysis stage produces the *requirements specification* phase product, using the statement of requirements as the main input. The aim of the requirements specification is to act as an overview of the desired system, in a structured form but using natural language.

NO STANDARD APPROACH

It should be noted that although the four methods for requirements acquisition are widely used, their results may be expressed differently. In addition, a feasibility study does not occur in all approaches to systems development. No agreement exists in terms of the contents of the statement of requirements for components that should be included or the level of detail required.

One of the reasons for this is that the methods used to analyse organizations and evaluate proposed projects involve management techniques and managers, rather than computer techniques and computer experts, and the study of these techniques has not, until recently, been undertaken from within the computer community.

Analysis

AIMS

The *analysis* phase has been influenced over the past few years by the conceptual modelling approach, and more recently by the object-oriented approach. Conceptual modelling aims are:

1. To model the desired system *precisely*.
2. To model on an *abstract level*, that is with no details concerning data representation or computer implementation.
3. To model *naturally*. This concept means that constructs in the modelling language are used which correspond on a one-to-one basis with user concepts of the components of their application.

To amplify these remarks, conceptual modelling aims to add more detail while retaining user terms and concepts (the user semantics), so that the description would be recognizable to a user. It differs from the previous phase as it uses more precise languages, some of which are mathematically based.

Conceptual modelling has also influenced the contents of the specification built in the analysis phase, which we define here as having three components: structure, process and rule. Each of these will now be briefly considered. We shall refer to the product of the analysis phase as the *specification*.

STRUCTURE COMPONENT

The structure component consists of entities, attributes and relationships, which are usually specified diagrammatically, and a widely used method is the entity-relationship model. The term structure is used as these objects are increasingly perceived as the focus of analysis, supporting other parts of the specification.

They are refined from the requirements specification, which, so as not to obscure its basic nature as a system overview, contains only the most important objects and properties. In addition, the normalization technique from the relational model may be used, especially where existing reports or screen formats, for example, are to be incorporated into the system.

RULE COMPONENT

Rules are restrictions in the organization that may be expressed in terms of restrictions on the entities, attributes and relationships in the structure component. For example, in Case Study 1, there is a restriction on the number of repairs that a net may undergo, and we model this by restricting the values of the number of repairs attribute of the net entity.

Rules are specified declaratively with the elements of the structure component to which they refer, using a precise language, which is usually based on logic or sets.

PROCESS COMPONENT

Processes are taken from the requirements specification and refined into more detail, down to the level of primitive processes operating on elements of the structural component. All events are identified. Process control structure is also modelled as well as the structural elements that are referenced.

There are a variety of forms for specifying processes, for example process decomposition diagrams for showing detailed process refinement and events, state transition diagrams for specifying the user interface, and data flow diagrams or entity life histories, showing processes that operate on an object and process sequence.

The object-oriented approach integrates the specification of processes, the objects on which they operate and rules.

Logical design

The aim of the *logical design* phase is to produce a design of the desired system that will serve as a basis for a computer implementation. Hence, we begin to place the emphasis not on producing a specification whose components correspond closely to user concepts (such as entity, process and so on) but on transforming the user-oriented components from the analysis phase into integrated components that incorporate factors such as non-redundancy of data, speed of access and suitability for acting as the basis of a computer program, which are important for a computer-based system.

There are two major tasks in logical design. Firstly, the specification from the analysis phase is transformed and, secondly, the human computer system is designed.

TRANSFORMATION OF ANALYSIS SPECIFICATION

Structure

The significant difference in this phase is that the structure component is now represented by data. We may also group several attributes together and represent them with one item of data. For example, a street name, a city name and a postcode may be grouped into a data item termed an address. There will often be several possible ways of representing relationships. We may also use normalization in this phase. We decide exactly what data types are required for representation, how many characters are required for each data item, and we design records and files or databases to store the data, taking into account the type of processes that will operate on the data.

Rule

Rules are integrated with the processes that operate on the constrained objects, and their form is transformed from declarative into procedural.

Process

Processes are now expressed in terms of operations on data. An abstract programming language form is often used here, such as structured English, JSP diagrams or action diagrams. In addition, data flow diagrams may be drawn showing the processes that occur, the data input to and output from each process, and data stores. Processes are grouped into logical units such as transactions and procedures; they are integrated into the human computer system and account is taken of whether processes are human or computer, on- or off-line.

DESIGNING THE HUMAN COMPUTER SYSTEM

Two levels of detail are normally considered here: firstly, the design of user procedures and, secondly, the design of the computer interface. User procedures consist of tasks and processes with which users will be directly involved, such as the data capture and transaction input processes from OMNIS, grouped into related units. In a transaction processing system, for example, a procedure will typically consist of tasks related to a particular event.

A major addition to processes is to consider the ways in which they are to be carried out. The types normally considered are (a) manual methods, that is filing, writing, information searching and so on, and (b) machine methods, which are mechanical (for example typing), and electronic (that is computer processes). Non-computer processes should be designed to be satisfactory for human use.

Computer interface design consists of the detail of processes and the objects on which these processes operate, and may involve considerations related to interaction style (screen and report layouts, human–computer dialogues), specifications of manual or mechanical operations, and off- or on-line processing. These processes are designed using the data objects defined earlier.

Physical design

INTRODUCTION

Physical design is the last of the design phases, and we may consider it as consisting of three components, hardware, software and the human computer system.

The logical design from the previous phase may be used as a basis for the design of a system consisting of many different types of hardware, software and human procedures. For example, we may decide that software will be written in BASIC, will execute on a small IBM PS/2 and, from the example in Case Study 1, will require boat captains to total the daily catch.

Alternatively, we may choose a fourth-generation language such as DBASE IV for the software, an Amdahl for the machine and let the harbour staff total the catch. The reasons influencing the choice will not be discussed here, but they will obviously take account of factors such as project estimated cost, process response time, storage capacity, number of users and so on. If a feasibility study has taken place in requirements determination, then it will need to be consulted. An obvious point is that the system components chosen, particularly the hardware and software, must be capable of working together.

The order in which we discuss the three components does not reflect the sequence in which they are considered at design time, as choices affect each other.

HARDWARE DESIGN

The hardware design consists of a description of the computers, storage devices, input/output devices and possibly networking devices required for the desired system. The hardware used for implementation and testing, and for operation may be different. This is often the case when a system is being built that will be sold to many different users who will run it in different hardware/software environments. The terms development and delivery environments are often used to distinguish the two.

Factors to be taken into account include: storage capacity and access times for storage devices, memory capacity and response time for processors and networks, cost of hardware, support for desired software, programmability of interface devices, quality and speed of printer required and so on.

SOFTWARE DESIGN

Software consists of the programs that run on the hardware. We shall include decisions about the physical design of data under this heading as the kind of data invariably affects the programs that process the data. Data is used by the programs in computer memory and 'persistent' data (data which has a life longer than the execution of a program) is stored on storage devices. It will be necessary to decide on the appropriate types of applications software, including languages and packages, as well as the systems software required to support the eventual system.

Process design

Traditionally, most applications software would be written in-house. However, there are now four options for process design:

1. For standard processes, we can buy packaged software. For example, accounting software is largely standardized. (An equivalent process is to contract out the software development to a third party, such as a software house or consultant.)
2. Use an application generator or fourth-generation language, such as Lotus 123.
3. Generate code using a CASE tool (see Chapter 11).
4. Write our own code.

If we write our own code, data structures are chosen for the representation of data in the programs, together with decisions as to the programming language required, for example COBOL, BASIC or ADA.

Detail is added to the logical design to take account of the data structures and data storage types chosen, and input and output data are designed for the processes. Issues such as execution speed and ease of maintainability of program code are considered, and processes are grouped into program modules.

Data design

The structural component, represented by data in the logical design phase, has decisions made for data storage and access on storage devices. Many different designs are possible and the feasibility of choices is determined.

These will partly depend on the decisions for processes discussed in the previous section. We may use manual means, or CASE tools, to design sequential or indexed sequential files, or databases. The maximum number of expected records will determine the file or database sizes required.

HUMAN COMPUTER SYSTEM DESIGN

The human computer system consists of hardware, software and processes. Processes will be carried out both by people and the hardware/software environment, and some processes will be manual processes that will not involve the computer.

Some of the design of this system has had to wait until the physical design phase, as the procedures are hardware and software dependent, specifying the activities to be followed when communicating with the computer, for example, detail of dialogue between computer and operator, procedures for starting up and shutting down the system and screen contents such as windows or colours.

On the periphery of the system, data capture processes may not involve computers at all, if they are off-line processes, for example, the data capture procedures in Case Study 1. Such processes may be assigned to certain departments, job roles or individuals.

Implementation and testing

MAJOR TASKS

The main output of the *implementation and testing* phase is a physical information system and not merely a design. Of course, the physical (and earlier) designs remain available for reference, as they form the specification.

The major tasks consist, firstly, of acquiring and integrating hardware, producing software, generating data for the files or databases and producing the human computer system. Secondly, the system is tested, and user comments are evaluated and perhaps used to redesign part of the system. Thirdly, in a stage termed post-implementation, the operation of the implemented system in the user organization is monitored closely for a limited period.

IMPLEMENTATION

Human computer system

The HCS processes require a clear definition for the humans who will carry out these processes. This definition should consist of natural language descriptions of procedures to be followed, and are typically written in manuals. They will cover details such as input preparation, how to request reports and so on. They are generally aimed on two levels, user management and users who will be concerned with functional information. Training for users will also be given.

Software and data

In-house implementation may consist of writing computer programs from the physical design, that is, producing source code or programming language statements. Alternatively, application generators may require forms to be filled in rather than program code to be written.

Programming development environments contain language-sensitive editors, compilers and interpreters, debuggers and other facilities such as version control to assist in this area. Data is generated for all entity, attribute and relationship instances and is used to populate the empty files and databases on the storage devices. This may be done manually or with the help of CASE tools.

TESTING

System testing, as the term is usually understood, refers to the demonstration of an implemented system's behaviour, with the aim of establishing that it behaves according to specification; that is it should behave in the way intended for the desired system.

System testing is rarely a structured procedure. However, a three-stage process may be identified. The first stage concerns individual programmers and designers who test individual units of the implemented system to their own satisfaction, supplying their own test data and procedures. In the second stage, the work of several people may be integrated and tested as a whole. Often, users may be asked to supply test data or even test procedures to provide a degree of objectivity.

It is only in the third and last stage that users are usually involved in evaluating a system. Typically, whole sub-systems will be demonstrated to users, usually with live data and performing tests designed by the users to assure them that the system does what they intended. The testing of the human computer system often takes place rather earlier than above, as factors such as the usability of the system for the intended system operators are crucial to eventual user acceptance.

It should be said that system testing is often a rather neglected task, with system developers anxious to obtain quick user acceptance and users unsure as to how thoroughly they should test the system.

POST-IMPLEMENTATION

Here the implemented system has been tested and is operating in the organization. It is often the case that where there is an existing system, both this and the new system run in parallel for a few months. This is in case teething troubles with the new system should cause it to break down.

In some cases, a post-implementation review may be carried out, after an interval such as six months or a year of operation. The procedure evaluates the behaviour of the implemented system against the original objectives, recommending changes if necessary to both the system and the process that developed it.

Maintenance

NEED FOR MAINTENANCE

Inevitably, some errors will exist in the system or the human computer system will require adjustment after practical experience. However, in addition, real-world changes will occur, which mean that objects or processes may be required to change also. Another source of change is technological, in that upgrades to hardware or the availability of improved software may necessitate radical system change. *Maintenance* is thus a re-execution of earlier phases of the process for the parts of the system that are evolving.

This phase has traditionally had little attention devoted to it. One reason may be that it is concerned with old applications, whereas new applications are more glamorous. Another reason may be that specialists find it hard to change old programs, often with no documentation, and prefer new applications with new technology.

However, recent work has emphasized the importance of maintenance, and various statistics have been used to underline the fact. For example, it is estimated that maintenance uses in excess of 50 per cent of systems development budgets, amounting to expenditure of over $30 billion per year, world-wide (Rock-Evans and Hales, 1990). This means that, over all systems, more resources are consumed on maintenance than on all the previous phases in systems development. A typical system takes between 1 and 2 years to develop and has a lifespan of over ten years before it is replaced. Thus, its maintenance life accounts for around 80 per cent of its lifespan.

Again, there are estimated to be about 75 billion lines of code in use, world-wide, some or all of which require maintenance attention. Of the estimated population of 10 million programmers by the year 2000, the great majority of these will be performing maintenance tasks.

TYPES OF MAINTENANCE ACTIVITY

Three basic types of maintenance activity have been identified:

1. *Corrective maintenance* is performed in response to processing, performance or implementation failures. It corresponds most closely to the traditional, narrow, view of maintenance whereby systems must be kept

up and running. It involves 'firefighting' in response to emergencies, such as program failure in the middle of execution, as well as more routine tasks, concerned with bringing code into conformity with specifications or standards.

2. *Adaptive maintenance* is performed in response to anticipated changes in the data or processing environments, for example payroll changes due to new legislation.

3. *Perfective maintenance* is performed to eliminate processing inefficiencies, enhance performance or improve maintainability. Providing user enhancements is the major portion of this task, keeping in step with the evolving needs of users.

Of the three types, perfective maintenance accounts for more than half the total effort.

PROBLEMS OF MAINTENANCE

Problems in performing maintenance activities have been found (Swanson and Beath, 1989) to depend on the following factors:

1. *User knowledge.* This factor includes seemingly insatiable demands from users for system changes, lack of understanding of current systems and unrealistic expectations.

2. *Programmer time availability.* This concerns the normally high programmer turnover rate and the preference of programmers for development rather than maintenance work.

3. *Product quality.* In many cases, the installed systems that are to be maintained have many errors or the documentation is inadequate.

4. *Programmer effectiveness.* This factor relates to the skills, knowledge of the installed systems and motivation of maintenance programmers.

REASONS FOR PROBLEMS

The third factor, *product quality*, is usually regarded as the most important, as this drives the other three. Poor product quality is due to:

1. Errors made during requirements and design phases.
2. Requirements changes occurring during development but not incorporated in the system.
3. Lack of, or inadequate, documentation of systems, necessitating detailed effort to understand often obscurely written programs.

The fact that perfective maintenance requires more effort than other types may conceal the fact that what computer management term 'providing user enhancements' may in fact be fixing errors or omissions made during development.

Although it is often assumed that maintenance tasks are analysis and design tasks, this is not the case. For example, an initial step is to check an existing program to see how it works. This involves system familiarity and diagnostic skills that are not required during systems development. In many cases, it may take longer to understand the program than it does to make a change.

SOLUTIONS

To alleviate maintenance problems it is suggested that maintenance should be separated from development, a maintenance career path for staff designed, more knowledge of maintenance (including statistics) should be accumulated and maintenance aids should be used.

Maintenance aids are methods or software tools that aim to improve productivity or quality of the maintenance activities. *Reverse engineering* tools have recently become available that analyse poorly documented systems and provide high-level descriptions, making the systems easier to maintain. For example, a type of data analysis may be performed on program code to show control flow and the flow of data. Another example uses file declarations to build an entity–relationship model.

Summary of the traditional approach

LINEAR MODEL

The traditional approach may be termed a *linear model*, as it regards phases, and stages within phases, as executing in a sequential fashion. The model rather oversimplifies actual practice, although it is useful as a framework within which the main activities may be discussed. We may summarize the advantages and disadvantages of the linear model as follows.

Advantages

1. The phases are clear-cut, with a beginning and end, making them easy to manage.
2. All the analysis and design is done before coding starts. If this were not the case, some parts of the system might not fit with others, requiring redesign or recoding.

Disadvantages

1. The assumption is that requirements are perfectly known before the project starts. However, requirements usually change as the project

progresses and as the user understands the implications of the requirements.
2. There is no guidance for understanding the organization in which the information system is to be situated. For all but the simplest systems, there is a danger that the wrong activities to assist will be chosen.

Most suitable use

The most suitable use is for projects with fixed requirements based on an existing system.

Dissatisfaction with the traditional approach

PROBLEMS

Although the process described above is well known, it is not ideal. Information systems development has been under pressure for some years to come up with an improved process that removes, or at least alleviates, some of the major problems that users experience in the development, operation and evolution of their systems.

Increasingly, development is taking place against a background of documented dissatisfaction with many aspects of the process. For example, a recent UK survey by the consultants KPMG Peat Marwick McLintock (*Computing*, 11 October 1990) found that '30 per cent of the UK's biggest computer projects were massively over budget, over time, and, if ever completed, fail to do the job they were meant to'.

More specifically, many users question the *productivity* of the process, feeling that development is too slow and expensive. Alternatively, they question the *quality* of the product, as they find that the eventual system does not do what they want. Another aspect of quality is that many users find that systems have been designed for computer experts, rather than organization users, and they question the usability of their system. In addition, many job changes may result from the introduction of a system, which, when fully known, may seriously affect the acceptability of the system to users or unions in the organization.

It is common for more effort to be spent during system evolution, in the phase often termed maintenance, than in the previous development phases. However, this phase has been neglected until recently, and the *maintainability* of systems is a key problem. How can existing systems, which are always growing in size and complexity, be easily changed to incorporate, for example, changes to old functions or the addition of new functions?

Another problem is *reliability*, where systems in safety-critical areas such as nuclear power stations and aeroplanes, as well as in organizations that carry out large financial transactions where errors are costly, are becoming increasingly dependent on computer-based information systems. Finally, *security* of systems is increasingly under threat from human agents (for example via hacking) or software agents, such as viruses. We will attempt to show later how improvements in information systems development are being made in an attempt to address these problems.

Quality and productivity are the two problems that are emphasized most at the moment, and we shall therefore examine their main causes in more detail.

QUALITY PROBLEM

A poor quality system is one that, for example, does not conform to its specification, does not have a positive effect on the organization or is not used by its intended users. The main reasons for poor quality are:

1. The information system may address the wrong problem, as it may not improve organizational efficiency or effectiveness, as discussed in Chapter 1. Alternatively, the system may conflict with organizational aims or strategies. In general, the wrong activities to assist are chosen, which may be due to lack of business knowledge or ignorance of wider organizational strategy.
2. Wider social or psychological factors may be neglected, such as the degree of decentralization or centralization of the organization or the extent to which the information system will be acceptable to or usable by its intended users. Systems that have low acceptability often fall into disuse.
3. The right activities may be identified, but errors may be made by users or analysts in analysing information needs. This is often due to the use of poor development techniques or the degree of complexity inherent in the development process.
4. The system may be developed for the wrong reasons, such as technology push from technical experts or political pull from ambitious managers. There are often problems with technocrats, keen to work with new hardware and software, as well as those who seek to extend their power or influence with a state of the art computer system, whether it is necessary for the organization or not.

The majority of these problems are due to the fact that the linear model does not pay enough attention to the nature of the relationship, discussed in Chapter 1, between the information system and the organization, and relies too heavily on analyst intuition, providing little guidance for the process.

PRODUCTIVITY PROBLEM

It is common for system development costs to be underestimated by a factor of two or more. This obviously affects the cost/benefit justifications made for the system in the feasibility study stage. It is also common for systems to be delivered later than planned, perhaps by a year or more. In this case, it is possible that the system may not be useful to the organization, as events in the market or environment may move so quickly that the system is out of date when delivered. Both these factors may cause system development to be cancelled or, if the system is delivered, never used.

An important factor causing poor productivity is *changing requirements.* The linear model assumes that requirements are known before the project begins. However, it is usually the case that information system requirements are changing continuously throughout the systems development process. If the specification of a system is always changing in this way then new work must be done and completed work redone, both factors causing costs to grow and delivery times to lengthen.

Reasons for changing requirements

Three main reasons why requirements change are:

1. Users are often not sure of their requirements at the beginning of a project, and they only discover what they really want as the project progresses and as they learn what they can realistically expect from a computer-based information system, with the existing budget, time-scale and development staff. There may also be conflict between multiple users.
2. Changes in external factors such as technology, legislation, the market or the political environment often change system requirements.
3. There may be implementation implications contained in requirements that are not feasible and are recognized only during implementation and testing.

Poor project control

A fourth reason for low productivity is that the estimation and tracking techniques used for project control may be inadequate:

4. Many projects have no means for their status to be measured at any given point, in relation to the work already completed or remaining to be done. In addition, at the start of a project, techniques for estimating the resources required are inadequate, based only on previous experience, and often give rise to over-optimistic predictions for delivery dates and costs. For example, it is common practice not to include maintenance costs in the cost estimates for a system.

In addition, the problem of low productivity takes place within the context of a shortage of experienced development staff. Problems of quality and productivity will interact, as, for example, the discovery during testing that a system is of poor quality may mean a delay while the system is changed, and changing requirements that are not taken into account result in a system of low quality.

IMPLICATIONS OF PROBLEMS FOR THE ORGANIZATION

It should be emphasized that there may be serious repercussions from these problems, such as a simple increase in costs, a write-off of all development costs for a system amounting perhaps to millions or tens of millions of pounds, poor staff morale owing to an unwanted system, missed commercial opportunity due to late delivery or a threat to organizational survival if a system contains many errors and existing customers are lost.

DIRECTIONS FOR IMPROVEMENTS

The main alternatives to the linear model that we shall discuss are based on the assumption that the above problems are best addressed by emphasizing the correct definition of requirements within the development process. The justification for this is as follows.

Figure 6.2 shows the resources estimated to be spent on each phase as a proportion of all non-maintenance phases (adapted from Boehm, 1981). These figures are averaged out from 63 projects undertaken in the United States, mostly in the real-time processing area. Note that the percentages refer to the total resources consumed and not the time taken.

Figure 6.3, based on surveys undertaken by de Marco, quoted in Finkelstein (1989), shows the sources of system errors and the relative cost of their correction. Figure 6.3(a) shows the causes of errors in systems, with errors in program coding contributing to only 7 per cent of errors found. The major

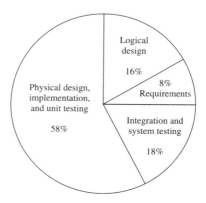

Figure 6.2   Resources spent on each phase as a proportion of all (non-maintenance) phases

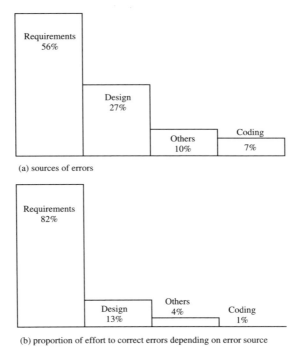

(a) sources of errors

(b) proportion of effort to correct errors depending on error source

Figure 6.3 Error detection and correction in maintenance (from Finkelstein, 1989)

problem is due to errors in the early, requirements phase (56 per cent). Comparable figures are also reported in Boehm, McClean and Urfrig (1975). Figure 6.3(b) shows the maintenance cost of eliminating errors, depending on the phase in which the errors were made. The coding errors (7 per cent) require 1 per cent of the cost, the design errors (27 per cent) require 13 per cent, while the requirements errors (56 per cent) require 82 per cent of the cost!

Most errors are thus introduced long before coding begins, and hence are the most expensive to eliminate, as they will clearly have a great impact on designs and software based upon them, all of which will have to be changed.

Linear model alternatives thus mostly aim to redistribute the proportion of effort shown in Fig. 6.2 away from later phases, such as implementation and testing, and into the early phases. However, they do not address all the problems discussed above, concentrating mainly on the third quality problem and the productivity problem. The formal methods approach is concerned with system reliability.

Alternative approaches to systems development

We shall discuss the *iterative, user validation, evolutionary* and *prototyping* approaches, which are all basically variants of the linear model. In addition, we shall look at the *formal methods* approach, which is rather different and relies upon formal specifications and program verification.

ITERATIVE APPROACH

An important modification to the linear model incorporates *iteration*, and we shall briefly explore the concept. The term iteration is applied to rather a wide set of activities. We may define it, within the context of models of the systems development process, as the process of performing a task in a phase more than once. It may be seen as the process whereby we go back over ground which we have covered before, repeating or reiterating that task. In practice, due to the interrelatedness of the specification components, we usually have to iterate many tasks at once. We shall discuss this topic from two viewpoints, iteration within a phase and iteration between phases.

Iteration within a phase

The discussion concerning the phase tasks of the traditional approach may have given the impression that when the first task finished, the second was begun, and so on. In other words, the tasks were totally ordered. In fact, it is common to have no particular order for tasks within a phase, and it is often necessary to iterate tasks before the phase product produced is satisfactory.

This applies particularly in requirements determination, which is not a phase that can be executed only once. It normally requires several repeated attempts, which refine the requirements obtained by analysing them and then checking them with the users, until a satisfactory requirements specification is obtained. This is partly because early requirements tend to be incomplete and inconsistent, and partly because requirements often change as the project progresses.

Study of a particular phase product in requirements analysis may reveal, for example, that a certain part is incomplete or that one part conflicts with another. In this case, the requirements acquisition task which obtains information must be repeated until the situation is clarified. A similar problem may occur in later phases where, for example, a data structure might be found to be incomplete, when the information required by all queries on that structure is cross-checked with the information in the structure. It is not possible to say how many times a particular task should be executed until it results in a satisfactory product, and there is no general model of the process that attempts to take account of this type of iteration.

Iteration between phases

Another implicit assumption of the linear model is that work only begins on the tasks of phase 2 when all the products of phase 1 have been completed, and so on. This is a management perspective of the process. However, we often need to repeat the work of earlier phases, particularly if we discover that a part of the evolving specification is incomplete and should have been captured earlier.

Figure 6.4 shows a model that takes account of iteration between phases. This model only assumes iteration to the previous phase. Variants of the model that allow iteration between any phase have been termed 'loopy linear' (Hawryszkiewycz, 1988). An example might be that an overlooked process comes to light in analysis, so we have to go back to requirements determination to redo the requirements specification. An early non-functional requirement concerning, for example, system performance, may not be possible, and it is not until physical design or testing that we find out.

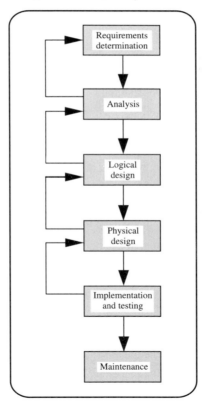

Figure 6.4 Iterative approach to systems development

Incorporating iteration into the systems development process is more a reflection of the real situation that occurs when systems are built, rather than a modification to the process itself. It aims to address the problem of incorrect requirements and changing requirements due to user uncertainty.

Advantages

1. The evolving system becomes closer to the desired system, as reworking is allowed.

Disadvantages

1. Uncontrolled iteration can cause a significant amount of wasted resources, owing to the need to redo earlier specifications. Where the span of iteration between phases is very great, for example, between the testing stage and analysis, then project progress and costs may be significantly affected.

Most suitable use

All types of project are suitable, but particularly those where the developers may be unfamiliar with either the application or the technology.

USER VALIDATION

The *user validation* approach aims to restrict iteration to the previous phase only, and involves the users in the *validation* activity of the development process. We may define validation as follows:

> Validation is an activity that checks that a product produced by the systems development process is correct, where the notion of correctness is that of correspondence to the user requirement.

An assumption of linear models is that users are rather passive 'consumers' in the process and that analysts and designers only 'empty' them of their knowledge. To counteract this, the user validation approach incorporates validation into the basic process by allowing phase products to be checked by the user. The general aim is to intercept any errors before work begins on the next phase, and so avoid iteration over many phases.

Of course designers do perform this activity in practice, but only on an informal basis. Validation has also tended to be restricted to the design and implementation and testing phases, where it is often termed verification or program verification. Structured techniques termed walkthroughs (Gane and Sarson, 1979), desk checking, chief programmer teams, and Fagan's inspec-

User difficulty in understanding specifications

There is a problem that runs through all of systems development, but which perhaps is experienced most acutely in the first phase. The problem is that there is a tension between a description of a system that is understandable to computer experts such as analysts and designers and a description that is meaningful and understandable to the user. This is probably based on the two different roles of a specification: the need to be a precise description of the desired system and the need to communicate the nature of that system to the eventual user of the system.

There are two aspects to this. Firstly, precise system descriptions by computer experts tend to use unfamiliar, computer-oriented languages. Secondly, a user tends to visualize a system as a physical system, that is in terms of processes executing, receiving inputs and producing outputs. A user finds it hard to tell from an abstract system description what the physical system will be like, even if the language is user oriented. Users feel happiest when they are validating an implemented system, so that they can experience its operation for themselves.

The danger of developing a system in abstract is that it may not be until the implementation and testing phase that users find out what the physical system is really like. If the users require major changes, then extensive and costly iteration must take place.

Features of prototyping

The linear model is modified by the prototyping approach, as shown in Fig. 6.7, where a prototyping cycle is taking place during the requirements determination and analysis phases, although it may be used in principle in any phase.

The approach provides users with an implemented system that they can try out, and that serves as a basis for improvements which can ideally be incorporated in the prototype and re-presented to the users. Hence, this avoids the problem discussed above of presenting only an abstract specification. The approach also takes into account the fact that many users do not in fact have a fixed requirement, as most models imply, and that experimentation with a real system enables users to better define, refine and communicate their requirements to the designers. Prototyping languages are also available which help to produce a prototype more quickly than conventional languages.

Advantages

1.   The availability of a real system, as opposed to a specification, allows for more effective validation in the early stages, addressing:

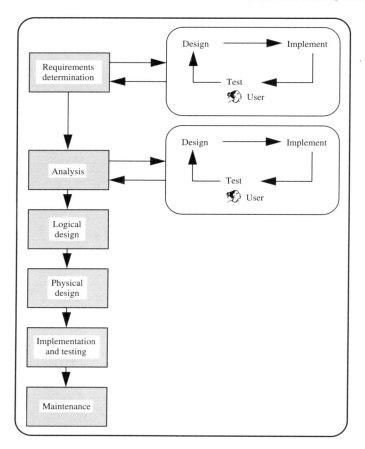

Figure 6.7 Prototyping modification to the linear model

(a) error correction by users
(b) definition of requirements, as it is easier to suggest changes based on experience of a real system, if the requirements are uncertain at the beginning
(c) incorporation of changes to an evolving requirement, due, for example, to changing environment or technology
2. Fast production of prototypes is possible using prototyping languages.

Disadvantages

1. It has been found expensive to produce prototypes, in terms of pro- grammer effort.

2.   Projects can be hard to control as users continually retune their require-
     ments; that is they spend a lot of time making changes that appear to be
     only 'cosmetic'.

Most suitable use

An early use of prototyping was only for naïve users with little or no
experience of systems development, where the requirements were likely to
change as users learnt their requirement. However, it is now used more
widely, typically for prototyping screen dialogues and input/output forms, as
it is increasingly recognized that validation is more effective with a real system
than with specifications.

An early claimed advantage of prototyping was the fact that it did not
produce documentation. Although this was always a doubtful advantage (for
example, how could a prototype be maintained if no documentation existed?), it was claimed in a context of linear manual development methods
which produced excessive documentation that quickly became out of date.
However, this has been improved by the recent trend to CASE tools, which
automate the storage and maintenance of the specification.

An improvement on the prototyping approach, termed *specification execution*, uses a specification language that can generate code to form a working
prototype of the desired system. This combines the benefits of checking an
actual system with the existence of a specification. The benefits of a
specification are emphasized in the next section on formal methods.

Another variation on the prototyping approach is to combine it with the
evolutionary approach discussed above, whereby prototyping languages are
used to produce parts of the implemented system for testing. An example of
this is the Systemscraft method (Crinnion, 1991).

Formal methods

The approaches to systems development described above all employ intuitive
methods towards the construction of a system. What we mean by this is that
we can never be sure that the software will work correctly, as the designers
are only relying upon their experience. The *formal methods* approach aims to
improve one aspect of this situation by attempting to demonstrate that a
program meets its specifications. It sets out to prove that a program works
exactly according to its specifications. However, the approach does not
guarantee that the specification meets the user requirements (Gehani and
McGettrick, 1986).

The approach is not widely used, as it is a relatively new development
requiring many hours work from highly trained specialists, and is therefore
expensive. Typical applications address the reliability problem and have been

in 'safety-critical' systems, such as nuclear power stations, aircraft, medical and car anti-lock braking systems, where reliability is crucial and human safety depends on software executing in a predictable manner. It has also been applied to situations where error is costly in financial, rather than in human, terms, such as financial transfer systems for banks. It may also be suitable for 'one-off' applications, such as launching nuclear missiles, if adequate simulations are not possible!

Because of its cost, the approach is only used to develop the most critical part of a system; it would not usually, for example, be applied to the production of user manuals or screen design. In practice, it is therefore usually combined with one of the 'intuitive' development processes previously discussed.

Figure 6.8 shows the initial phases of a formal methods approach and the phase products. Its typical phases are as follows.

PHASE 1 — REQUIREMENTS ANALYSIS

This phase produces a requirements specification, which is an informal description of desired system behaviour. It details the tasks that the system should perform, as well as any restrictions on time, space or other resource utilization for accomplishing these tasks.

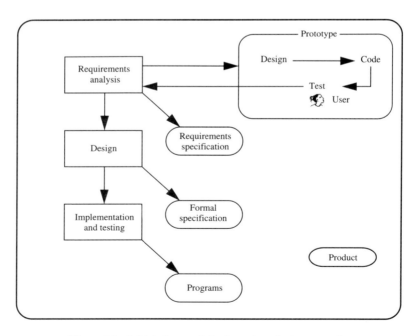

Figure 6.8   Initial phases of the formal methods approach

At the present time, this argument appears plausible, but it is difficult to set up a proper study to investigate whether in fact the predicted maintenance savings do occur.

Discussion questions

1. Draw a diagram summarizing the main phase tasks and products of the traditional approach to systems development. Which of the products do you think are suitable for validating with users?
2. Contrast the aims of the requirements determination, the analysis and the logical design phases.
3. Why do you think, as suggested by Fig. 6.3, that most errors are made in the requirements phase?
4. If you were a project manager, how do you think you would try to control iteration in your systems development project?
5. What are the reasons for calling the model in Fig. 6.5 user validation?
6. What are the significant differences between validation and prototyping?
7. In which area of Fig. 6.3 do you think that the formal methods approach is likely to contribute most to systems development?

References

Boehm, B. W. (1981) *Software Engineering Economics*, Prentice-Hall, Englewood Cliffs, NJ.

Boehm, P. W., R. K. McClean and D. B. Urfrig (1975) 'Some experience with automated aids to the design of large-scale reliable software', *IEEE Transactions on Software Engineering*, vol. SE-1, no. 1, pp. 125–133.

Crinnion, J. (1991) *Evolutionary Systems Development: A Practical Guide to the Use of Prototyping within a Structured Systems Methodology*, Pitman, London.

Davis, G. B. (1974) *Management Information Systems: Conceptual Foundations, Structure and Development*, McGraw-Hill, London.

de Marco, T. (1982) *Software Systems Development*, Yourdon Press, New York.

Fagan, M. E. (1976) 'Design and code inspections to reduce errors in program development', *IBM Systems Journal*, vol. 15, no. 3, pp. 182–211.

Finkelstein, C. (1989) *An Introduction to Information Engineering*, Addison-Wesley, Wokingham.

Gane, C. and T. Sarson (1979) *Structured Systems Analysis*, Prentice-Hall, Englewood Cliffs, NJ.

Gehani, N. and A. McGettrick (1986) *Software Specification Techniques*, Addison-Wesley, Wokingham.

Gilb, T. (1988) *Principles of Software Engineering Management*, Addison-Wesley, Wokingham.

Hawryszkiewycz, I. T. (1988) *Introduction to Systems Analysis and Design*, Prentice-Hall, London.

KPMG (1990) *Runaway Computer Systems — A Business Issue for the 1990s*. Available from KPMG Peat Marwick McLintock, 1 Puddle Dock, Blackfriars, London EC4V 3PD.

Longworth, G. (1989) *Getting the System You Want: A User's Guide to SSADM*, NCC Publications, Manchester.

Rock-Evans, R. and K. Hales (1990) *Reverse Engineering: Markets, Methods and Tools*, Ovum Ltd., 7 Rathbone Street, London W1P 1AF.

Swanson, E. B. and C. M. Beath (1989) *Maintaining Information Systems in Organizations*, Wiley, Chichester.

7
Requirements determination

Introduction

Requirements determination is the least well-defined phase in the systems development process. One reason for this is that it is the least technical, and therefore the most organization dependent. This means that procedures and products are likely to vary greatly from organization to organization. In addition, many organizations are realizing the advantages that result from strategic planning, discussed in Chapter 14, which analyses the information needs of an organization and produces an information systems plan to match. If a strategic plan exists, then some of the early activities described in this chapter will have already been done.

Another reason lies in the nature of the requirements themselves, which, as discussed in the previous chapter, are by no means as clearly defined and fixed as is commonly supposed. Requirements frequently change once systems development is under way, for a variety of reasons, and, in addition, it is often difficult to be sure that the system whose requirements are being obtained is the best system. Different ways of coping with these problems cause differences between organizations in the way in which requirements determination is carried out.

Four stages frequently found in requirements determination are:

1. *Problem definition*
2. *Feasibility study*
3. *Requirements acquisition*
4. *Requirements analysis*

We will use a case study, based on Eurobells △, to demonstrate the use of the interview method as well as the iterative nature of requirements acquisition and to illustrate the application of the OMNIS model to requirements analysis.

Problem definition

REASONS FOR A SYSTEM

The original idea for a desired system often originates in a *system proposal*, which is typically an informal document concerned only with system scope and its justification. A system proposal may originate to achieve one or more objectives:

1. To solve a problem
2. To take advantage of an opportunity
3. To respond to a directive

To illustrate objective 1, in Case Study 1 in Chapter 4 a rather informal manual system may have existed for some time, recording the customers who buy fish on a daily basis. This has meant laboriously adding daily totals (many of which were incomplete and missing) to obtain monthly totals for individual customers. Also, daily totals were often exaggerated and incomplete, and it was felt that a computer-based system would help accuracy and provide information not available before.

Objective 2 concerns expanding or improving organizational performance, for example offering customers access to organizational databases so that they get a better service. An example of objective 3 is where legislation such as the Data Protection Act may necessitate changes to the way customer information is stored or accessed.

These are rather abstract and high-level objectives, and a proposed system should normally describe how it will contribute towards the efficiency or effectiveness of the organization. Projects may be undertaken for five general reasons (Senn, 1989), which we may classify as follows:

Efficiency improvements:
1. *Capability*. This concerns an increase in processing speed, volume or faster information retrieval, compared to manual equivalents.
2. *Communication*. This is concerned with communication both within and between organizations, mainly to increase the transmission speed of messages, for example EDI or electronic funds transfer (EFT). Also covered is integration of business areas.

Effectiveness improvements:
3. *Control*. This aims to improve management controlling functions, and includes procedures for greater accuracy and consistency, and improved security.
4. *Cost*. This covers cost monitoring for labour, goods or overheads and their relationship to departments or individuals.
5. *Competitive advantage*. Factors here are locking in customers, locking out competitors, improving arrangements with suppliers and developing new products.

Projects concerned with communication may contribute towards effectiveness as well as efficiency.

A system proposal usually originates from a section manager, senior executive or analyst, and should not take any longer than a few days to compile. After the problem definition report has been prepared, which is generally in a standard organizational format, it has to be accepted by a committee in order for it to pass to the feasibility study stage, which is an exercise that consumes more resources than problem definition.

Project selection committees are usually one of three types: (a) steering committee, which usually contains a majority of senior departmental managers, (b) information systems committee, whose membership is usually drawn from the computer department, (c) user group committee, a departmental committee consisting mainly of users together with department computer specialists. If the organization has a strategic plan, as mentioned in the introduction, then the relationship of the project to that plan should be determined here.

REPORT

The form and content of the problem definition report will vary between organizations, but its contents are typically:

1. A statement of the problem to be solved and its significance, or a justification for the system.
2. The suggested solution, how an information system will help and the benefits expected.
3. The scope of the desired system.

The system may be a simple upgrade to an existing system or a multimillion pound organization-wide system involving a major change in hardware supplier. The detail required in reports in these early stages will obviously depend on the size of the system.

Feasibility study

The committee will have narrowed down the number of potential projects to pass a proposal to the *feasibility study* stage. The aim here is to evaluate the feasibility of the proposal, involving an analysis of the problem and the determination of the best solution within the context of the organizational situation. It is usual for several alternative solutions to be prepared from a proposal, usually ranging in the scope of their functionality, that is in the size of the proposed system.

This stage will prepare a high level set of non-functional, as well as functional, requirements for each of the alternative systems. Functional requirements concern the functions of the system, that is what the system will do. Non-functional requirements concern resource restrictions on the system, such as the maximum number of users and factors such as future expandability, security and reliability.

To pass this stage and to go through to systems development, a proposal must demonstrate (Kendall and Kendall, 1988):

1. That it will help attain organization objectives
2. Economic feasibility
3. Technical feasibility
4. Operational feasibility

ORGANIZATION OBJECTIVES

A topic that may be investigated here concerns the nature of the problem asserted in the problem definition. Has the problem been defined correctly? Another issue that may arise concerns the motivation for the proposal, which may be due to internal organizational power battles, prestige of the computer department or similar reasons.

If the organization has a strategic plan, the proposal should be checked for conformity to this. For example, an organization may have a five-year plan for IT development, which has an overall budget and has, as a priority, horizontal integration over different departments, such as, for example, sales, marketing and manufacturing. Another element in a strategy might be to give priority to systems for automating areas that are still largely manual. The priority of the system as a whole, and different parts of it, should be described.

ECONOMIC FEASIBILITY

The aim here is to estimate the costs required for alternative systems and set them against the expected benefits. Obviously, a successful system will aim to benefit the organization! The types of alternatives that are frequently considered are the placing of the manual/computer boundary, as some tasks may benefit more than others from being computerized, and non-functional requirements such as the time delay between the real world and different parts of the information system, which may be batch, on-line or real-time. Such decisions will also have an impact on technical feasibility, discussed below.

The systems also have their costs estimated in terms of the basic resources of money, people and time. For example, the following must be costed:

1. Systems development, involving, for example, analysts and programmers for systems development, or consultancy costs
2. User time for requirements, testing and training
3. Hardware and software acquisition

To set against the costs, the expected benefits should be quantified, for example, reduced costs, improved customer service or a predicted increase in orders.

This area is rather unsatisfactorily served at the moment, because apart from transaction processing systems which may result in quantifiable efficiency increases, it is very difficult to predict with any certainty whether a given system will in fact benefit the organization, as many factors are involved in system success. This point will be developed later in Chapter 12. In addition, on the costs side, the cost for maintenance (which may be several times the cost of developing the system) is rarely included.

TECHNICAL FEASIBILITY

Technical feasibility is concerned with determining whether a solution can be implemented on existing technology. If it can, then current technical resources may require upgrading or adding to. Some solutions might require the use of very new equipment or software that have not been integrated before.

Non-functional requirements are taken into consideration here, such as on- or off-line processing, maximum response times for user–computer interaction, estimated hourly or daily frequency of transactions, maximum record and file sizes for storage, networking loads and typical numbers of users at one time. In addition, requirements for system expandability, security, data archiving and reliability are considered.

OPERATIONAL FEASIBILITY

This investigates factors such as the likely reaction of organizational employees and union representatives to job and other proposed organizational changes. The main aim is to assess whether the solution will operate and be used after installation. For example, if users are happy with their current system and see no need to change, there might be a high degree of resistance to the new proposal.

Relevant factors here concern whether the solution has general management support and whether or not proposed system users have been involved with the development of the proposal. In general, a project without management support is unlikely to succeed, and projects with a high degree of user involvement are more likely to be acceptable to the users.

OUTPUT FROM THE FEASIBILITY STUDY

A problem of the feasibility study is that many questions cannot be properly answered at such an early stage of the project. In addition, a key factor in successful development which is not covered is the availability of sufficiently experienced systems development staff. In general, the study often adopts a rough and ready approach whereby it rejects proposals that do not fit in with the organization strategy, that appear technically impossible or improbable, where the cost/benefit equation looks wrong or where no user seems to want the system.

The study is typically carried out by analysts, who interview proposed users and review documentation; they will produce a report with recommendations. The report should contain a description of the functional requirements, as well as the important non-functional requirements, for the proposed system, and should cover the four points concerned with organizational objectives and feasibility discussed above. A recommendation might be to commission software from a software house, in preference to in-house systems development, for cost reasons. A recommendation with an operational reservation might be to install a proposed system in two stages, a pilot system followed by the full system.

PROJECT PLAN

For successful proposals, a plan for the systems development project, including a budget, time-scale and an estimate of different kinds of personnel required, will usually be based on the feasibility study. Project planning is a wide topic, which we cannot develop here, but an important element comprises a set of detailed activities for the project, usually broken down into a hierarchy. For example, the top level will often be in terms of the sequence and duration of the systems development phases that will be required. These may vary considerably from project to project as, for example, prototyping or incremental development may be used. Major deliverables will be identified for each phase and the extent of validation or quality control will usually be set.

The next level down will refine each phase into a set of ideal activities, based upon the resources (financial costs, man-days, elapsed time) available. Budget limitations are crucial here, especially when considering project team costs such as salaries or fees. Diagrammatic aids may be used, such as the Gantt chart, which is a bar chart showing the duration scheduled for activites and whether or not they have been completed. The PERT diagram is used for showing the sequence of activities and their scheduled duration, and is also used to derive the critical path of those activities whose completion is critical to on-time project completion.

Requirements acquisition

When the proposed system has been approved as a result of the feasibility study, the detailed requirements of the users are then acquired. We shall discuss the four traditional *requirements acquisition* methods and then present a case study resulting in the production of the statement of requirements.

production cycle and for when an order was received. A diagram was also attached. She began to read the background material, shown in Fig. 7.1 and Tables 7.1 to 7.3.

THE FIRST INTERVIEW AT EUROBELLS △

Wednesday 14 October, 11.05 a.m. Ken Campana's office

'Yes, you're right, the order we processed last week had to await manufacture. In the ideal situation of course there will be sufficient finished products in stock to meet any order. It is necessary for speedy dispatch to keep the stock levels for each product as high as possible.'

Orelie Gallo was having her first interview with Ken Campana at Eurobells △. She had assimilated the departmental description and had decided to use the interview method with Ken Campana to find out more about the problems in the stock room. She had scheduled the 45-minute interview yesterday afternoon.

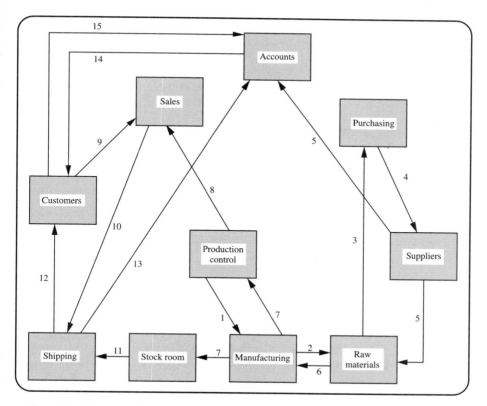

Figure 7.1 Sequence of processes through Eurobells △ departments on receipt of an order

Table 7.1   Eurobells △ feasibility study report — stock room

Objectives	Improved efficiency of some or all stock room activities so that stock levels are maintained at a minimum level
Technical feasibility	Any computer solution must be compatible with current Eurobells △ hardware and software strategy
Operational feasibility	Company policy is that there should be no staff redundancies
Economic feasibility	The proposed system budget is £150K capital and £20K recurrent annually

Table 7.2   Description of Eurobells △ departments

The Eurobells △ organization relies for its income on the process involved in making its bells and clappers. The functions of the various departments are:

Accounts	Deal with Eurobells △ finances including customer and supplier accounts, invoices, staff payroll
Sales	Have direct contact with customers through salespeople. They are the customer's link to the company
Purchasing	This is the department that buys raw materials from the suppliers. It is their job to buy as efficiently as possible
Manufacturing	The department that turns raw materials into bells and clappers, and assembles them into products
Production control	It has information regarding current stock levels, orders from customers, staffing figures in manufacturing and so on. They tell manufacturing what and when to manufacture
Stock room	A store room where products and raw materials are kept prior to shipping and manufacture
Shipping	This department packs the products (bells, clappers) into containers and sends to customers by road, rail, sea or air

'But there are two main problems with this approach, aren't there?' pointed out Orelie.

Ken looked apprehensive, then relaxed. 'You're right there!' he shouted. 'First, there's too much space required. That raises our costs — heat, light, we have to rent space sometimes. Secondly,' he pointed at the laptop on his desk, 'it's costing the company a fortune! Look at my estimate of the value of the current stock. There's just too much capital tied up in unsold goods.'

Orelie joined Ken behind the desk and examined the figures on the screen. She made some notes. 'So,' Ken continued, 'we want to keep stock levels as

Table 7.3 Sequence of processes through Eurobells △ departments

The sequence of processes through these departments for the production cycle and when a customer order is received is described below (see Fig. 7.1 also):

1. Production control tell manufacturing to make a quantity of bells and clappers.
2. Manufacturing tell the stock room (raw materials) the raw bronze it requires.
3. There is a shortage of raw bronze so the stock room (raw materials) tells purchasing (note that production control may know and may tell purchasing themselves).
4. Purchasing order the necessary raw materials from suppliers, having chosen the best supplier in terms of cost and delivery time.
5. Suppliers send the raw materials to the stock room (raw materials) and the invoice to accounts.
6. The stock room passes the raw materials to manufacturing.
7. Manufacturing makes the bells and clappers, assembles them into products, sends the products to the stock room and tells production control what products were made.
8. Production control tell sales what products are available.
9. The customer orders 5 tenor bells and 20 soprano bells. Sales receive the order and check:
    (a) Does the customer do business with us?
    (b) Is this customer's credit OK?
10. Sales tell shipping that the products are in the stock room.
11. Shipping obtain the products from the stock room and pack them into containers.
12. The products are sent to the customer by shipping.
13. Shipping tell accounts.
14. Accounts produce an invoice and send it to the customer.
15. The customer sends money to accounts.

low as possible to minimize the problems. However, as I guess you also know, keeping very low stock levels causes delays in filling orders and possible loss of customers.'

'That's right,' Orelie agreed. 'Each company therefore has to decide on a balance between very high stock and very low stock. What can happen is that a company can choose a stock level for each product which is a danger level. That's the level below which the stock should not be allowed to drop. But it can be complicated,' she continued, 'as the danger level will vary from product to product, depending on its popularity.'

'And in our field,' chimed in Ken, 'there's also the problem of seasonal factors. Christmas and Easter are our biggest times, and so the danger level will vary throughout the year.'

'What sort of control system do you have right now?' Orelie asked.

Ken looked defiant and then he laughed resoundingly. 'Well,' he said, 'it's like this. Come over to the window.' Orelie walked over to the open casement. 'What do you see?' asked Ken.

'There's a lot of packing cases,' she exclaimed.

'Those are the bells and clappers that we're not sure if anyone wants because we don't know what's in them,' Ken commented sarcastically.

'What are those teams of people doing with notebooks?' Orelie wanted to know.

'That's your control system,' Ken laughed. 'The problem, right, is the problem of monitoring stock levels in order that stock lows and highs can be highlighted,' Ken stated, staring at the shuffling teams. 'We want to keep an up-to-date view of stock levels of individual products. Well, the way it's done here is that there's meant to be a periodic stock check by clerks, who traipse around the store room counting products. The only thing is that it's a full-time job now! Our problem, and I hope you can help us here, Orelie,' Ken suddenly became more business-like, 'is that the current control system, if we may grace it with such a term' (here Orelie grinned), 'is that it's (a) time consuming, (b) boring, (c) useless for monitoring theft or loss as it's never complete, (d) prone to error and (e) always going to produce out of date figures.' Ken pointed at his right hand with its fingers and thumb out-stretched. 'Those are my five problems!' he complained, slumping back in to his chair dejectedly.

Orelie stood up. 'Thank you, Mr Campana, for an interesting interview. I think I've understood most of the problem. Let me fix another interview with you for Friday. I think I'll have a first proposal for you then.'

'I'll look forward to that, Orelie,' said Ken. 'Just remember, though, we don't want to lay off any staff. We want to keep on good terms with our people.'

Thursday 15 October, 3.40 p.m. Spiel-Jouet-Ludus offices

'Hi Orelie.'

'Oh, hi,' smiled Orelie.

Martine was her closest colleague at work. 'So, anything new? Hey, what's that?' she was quizzically examining Orelie's notepad contents.

'Don't touch it, it'll probably fall to dust,' Orelie joked.

'Museum piece, is it?' Martine called from the staircase.

Orelie looked at the proposal she was preparing for Eurobells △. She had the uneasy feeling that although she had covered the present situation in Eurobells △ she did not really know what sort of system Ken Campana had in mind. However, she at least knew that it only concerned the stock room.

The statement of requirements for Eurobells △ may be seen in Table 7.4.

Table 7.4   Statement of requirements — Eurobells △ stock control system

---

Current situation
Only an informal, manual information system exists. Staff have no clear procedures and no set way of recording data. It is not clear what data to record.

Proposed system

1. *Stock file.* A computer stock file is kept, altered daily, giving current stock levels for the different kinds of products and raw materials. The stock file will contain the following information for each item type:
   (a) Product/raw material code
   (b) Stock level
   (c) Danger level
   (d) Selling price
   (e) Cost price
2. *Stock in.* This covers raw materials as well as products and comes from one of two document types:
   (a) Suppliers supply document
   (b) Manufacturing completed worksheets
3. *Stock out.* Shipping ask for products by way of a shipping note. Raw materials are requested by manufacturing using a materials request form. For stock in and out, the documents contain source data which will be used to update the stock file. Keyed input data will be:
   Product/raw material code, quantity (+ or −), date
4. *At end of day.* These documents can be batched together and keyed into the computer. To give totals we use the formula:
   Current stock level = old stock level + additions from manufacturing and
                         suppliers − removals from shipping and manufacturing
5. *Reorder.* The system has software to report automatically when danger levels of products are reached. A stock reorder level field for each product is to be introduced. When this level is reached, the computer can print a report showing the items to reorder.
6. *Data collection.* An initial data collection phase is required to create the stock file on the computer.

Requirements analysis

In the *requirements analysis* stage, the narrative statement of requirements is analysed into a more structured description, usually termed the *requirements specification*. The advantage of such a description is that it can be more easily checked for missing or contradictory requirements (often referred to as the properties of incompleteness and inconsistency) by the analyst, and can be presented more systematically to the user for validation.

To do this, it is necessary to have a model into which the proposed system can be 'fitted', and in this section we show how the OMNIS model, discussed in Chapter 4, is applied to the statement of requirements to produce the requirements specification. More traditional methods tend to use computer-oriented analysis and specification techniques, such as data flow diagrams, which show only a partial view of the specification.

Case Study 3 — Eurobells △ stock control system

The statement of requirements requests a control system with a computer-based stock control information system. In addition, the management system activity of stock reorder, previously a manual activity, is now to become a computer activity.

Figure 7.2 shows we have simplified the stock control system to consider only four main 'production' processes, as the stock room does not store intermediate states of raw material, produced by the melt or mix processes, and nor does it store unassembled bells or clappers, produced by the cast bell and tune processes. Only raw materials and products are stored in the stock room. A product may be either an assembled bell and clapper, or a clapper.

REQUIREMENTS SPECIFICATION

SIMULATION SYSTEM

Information base

Object	Properties
PRODUCT-LEVEL	Product-code, stock-level, danger-level, price, reorder-level, reorder-quantity
RAW MATERIAL	Material-code, stock-level, danger-level, price, reorder-level, reorder-quantity

Rule base

stock-level ≥ danger-level

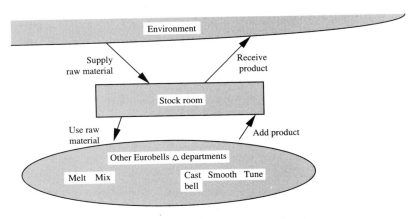

Figure 7.2   Schematic of stock control system

Process

Process	Refinement
1. Supply raw material	Increase stock-level of raw material
2. Use raw material	Decrease stock-level of raw material
	Initiate message system process 2
3. Add product	Increase stock-level of product-level
4. Receive product	Decrease stock-level of product-level
	Initiate message system process 1
5. Create, modify, delete	Product-level, raw material

MESSAGE SYSTEM

1. Check product-level stock-levels.
2. Check raw material stock-levels.

For the above processes, generate a list of codes and reorder quantities where stock-level < reorder-level.

3. Allow for reports on stock-level for all products and raw materials.

HUMAN COMPUTER SYSTEM

Input system

- *Data capture system.* Use copies of supplier documents, worksheets, material requests and shipping notes as source data. Batch until day end.
- *Transaction input system*

Event	Process
1. At day end	Use batched information to initiate simulation system processes 1 to 4
2. On request	Initiate simulation system process 5

Output system

1. MES processes 1 and 2 started by simulation system processes.
2. MES process 3 to be run at day end.

MANAGEMENT SYSTEM

Obtain the stock reorder list and send it to production control.

Functional and non-functional relationships

A *functional* relationship exists between two entities A and B where *any* instance of entity A may associate with *not more than one* instance of entity B at a time *t*. This may be seen in the top diagram of Fig. 8.5.

In contrast, a *non-functional* relationship between two entities A and B exists where *any* instance of entity A may associate with *more than one* instance of entity B at a time *t*. This may be seen in the bottom diagram of Fig. 8.5.

Three types of cardinality

Relationship cardinality concerns the functionality of both of the relationship directions. There are three types, 1 : 1 (*one to one*), 1 : *n* (*one to many*) and its inverse *n* : 1 (*many to one*), and *m* : *n* (*many to many*):

- *One to one.* Both directions of the relationship are functional. Therefore, for entities A and B, at a time *t*:

    One entity instance A associates with only one entity instance B.
  and
    One entity instance B associates with only one entity instance A.

  Figure 8.6 illustrates a 1 : 1 relationship between entities A and B, for example, a relationship between entities customer and account, modelling the fact that a customer has one account and an account is for only one customer.
- *One to many.* Only one of the directions of the relationship is functional here. For example, if A → B is 1 : *n*, then at a time *t*:

    One entity instance A associates with many entity instances B.
  but
    One entity instance B associates with only one entity instance A.

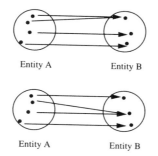

Entity A   Entity B

Entity A   Entity B

Figure 8.5 Functional and non-functional relationships between entities A and B

Entity A   Entity B

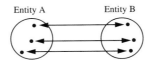

Figure 8.6   1:1 relationship between entities A and B

Figure 8.7 illustrates a 1 : *n* relationship between A and B in the A → B direction. A → B is non-functional (a1 → b1 and b2), but note that the inverse relationship, B → A, is *n* : 1 and is functional (b1 → a1, b2 → a1, b3 → a2, b4 → a3).

Note that not all instances of A have to be associated with more than one instance of B. As long as the possibility exists that any instance of A can associate with more than one instance of B, then the relationship is 1 : *n*.

A 1 : *n* relationship between two entities, for example customer → product, models the situation where there is a hierarchy that is a tree. Considering customer and product as parent and child respectively, then any given child (product) instance may have only one parent instance (customer).

- *Many to many.* Neither of the relationship directions is functional here. For entities A and B, at a time *t*:

    One entity instance A associates with many entity instances B.
and
    One entity instance B associates with many entity instances A.

Figure 8.8 shows an *m* : *n* relationship between entities A and B. There is no direction to specify because an *m* : *n* relationship is equivalent to both a 1 : *n* A → B and a 1 : *n* B → A relationship.

An *m* : *n* relationship between two entities, for example supplier and raw material, models the situation where there is a hierarchy that is a network and not a tree. Considering supplier and raw material as parent and child respectively, then any given child instance may associate with more than one parent instance, as a sort of raw material may be supplied by many suppliers, as well as a supplier supplying many sorts of raw material.

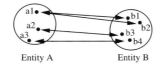

Entity A   Entity B

Figure 8.7   1 : *n* relationship between entities A and B (A → B)

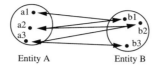

Figure 8.8   *m* : *n* relationship between entities A and B

DIAGRAMMATIC REPRESENTATION OF CARDINALITY

Relationship cardinality may be shown by a variety of methods. The one chosen here is the 'crow's foot' symbol, the absence or presence of which is used as shown in Fig. 8.9. Figure 8.10 shows an entity model incorporating the examples given above.

Determining relationship cardinality

* *Functionality determination.* When deciding on the functionality of a relationship between two entities, we must be careful to choose a value that will be true at *any time* in the life of the entity instances. For example, by simply inspecting the population diagram in Fig. 8.3 at a given point in

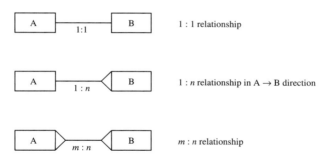

Figure 8.9   Entity model symbols for relationship cardinality

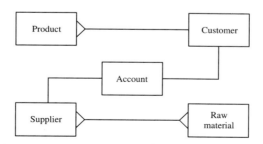

Figure 8.10   Example entity model for relationship cardinality

time for the customer and product relationship, one might infer that customers may buy only one product. However, this might just be the situation at that point in time and would not accurately reflect the fact that a customer instance may associate with more than one instance of product.

- *Meaning of 'many' as in '1 to many'*. Be careful that you do not equate 'many' with 'all'. It only means 'more than one'.
- *Symmetric and asymmetric relationships*. Note that 1 : 1 and $m : n$ relationships are symmetric, but that 1 : n relationships are not. For example, for the non-functional relationship customer → product, the inverse relationship product → customer is functional.
- *Fixed cardinality*. A relationship can be fixed cardinality, where the number of instances of entity B that may associate with an instance of entity A is restricted. For example, if a customer could buy a maximum of four products, the cardinality of the relationship customer → product would be 1 : 4.

Attribute

DESCRIPTION

This type of component is an entity descriptor, and an attribute is usually considered in relation to the entity it describes. For example:

Product   Product number, type, price
Customer   Name, address, telephone number, fax number

Product and customer are entities, whereas product number, address and so on are attributes that describe entities. For example, product is described by its price.

An instance of an attribute is termed an attribute value. For example, the attribute (product) type has two values, bell and clapper. As for other kinds of instance, attribute values are known only at run time.

ENTITY–ATTRIBUTE RELATIONSHIPS

As for entity–entity relationships, role name and relationship cardinality apply.

Relationship cardinality

- *One to one*. An attribute in this type of relationship is termed the entity *identifier*, as each attribute value uniquely identifies an entity instance.

materials R1 and R2 may be associated to many different projects. This means that we cannot know, for raw materials R1 and R2, the project(s) to which they were supplied by supplier S1.

We may model these *n-ary situations* by modelling the *n*-ary associations as an entity. We create a new entity, formed out of the *n* (in this case, three) entities concerned, forming an entity we name SRP. This is then associated to the other entities via binary associations, as in Fig. 8.16. Supplier S1 is now associated to instances of SRP, each instance of which is associated to an instance of raw material and an instance of project which receives that raw material from that supplier.

The entity SRP is rather artificial, and an example of a more natural entity that models an *n*-ary situation is the entity complaint, where a customer complains about a product sold by a salesperson. In this case, Fig. 8.17 shows the modelling.

Eurobells △ entity model

Figure 8.18 shows a partial entity model of the Eurobells △ production system. The product entity, which is the final object produced by assembly, has two entity subsets, bell-product and clapper-product, which are formed by clustering on product-type values.

Product and raw material entities have relationships to the attributes as shown, and, in addition, they have relationships with other entities. Product has relationships with customer (the customer that buys the product), bell

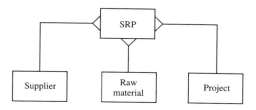

Figure 8.16   *N*-ary relationship modelling of supplier, raw material and project

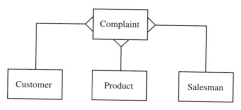

Figure 8.17   *N*-ary relationship modelling of a complaint concerning customer, product and salesperson

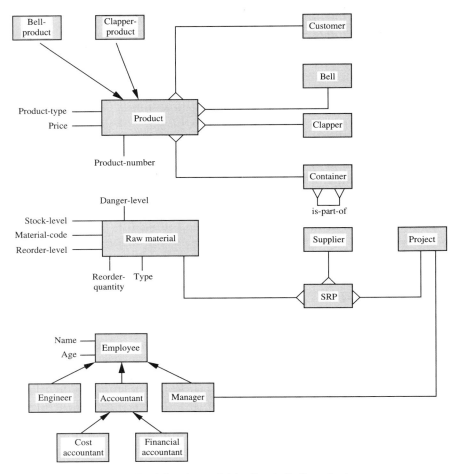

Figure 8.18 Partial entity model for Eurobells △ production system

(the bell that constitutes the product), clapper (the clapper that constitutes the product) and container (the container used for the product). These relationships are all $n : 1$, from the product direction, so a product is made up of only one bell, clapper or container.

Raw material, supplier and project entities are shown in an n-ary situation, and container has an $m : n$ relationship with itself, reflecting the fact that a container may be made up of many containers. Finally, the employee generalization hierarchy is shown.

Abstraction

When we build an entity model, we make use of three abstraction operations: *classification*, *generalization* and *aggregation*. An abstraction operation is an operation we perform on a group of objects to reduce the amount of detail that is visible, that is to reduce their complexity.

CLASSIFICATION

The operation of grouping similar objects into classes and sets, discussed above in the entity section, is termed classification, and an important part of the operation is to give the set a name, for example *employee*. The name corresponds to the class name in the real world, denoting a class of objects, and it usually attempts to express the difference between this and other classes. The operation reduces the complexity of many employees to just the employee class, so making the distinction between class and instance. Although the user often knows his or her basic entities, analysts often use this operation to build abstract entities during the modelling process, as in the example of *n*-ary relationships. In addition, decisions must often be made concerning the level of granularity that is required concerning sets of objects, for example concerning the raw material entity.

GENERALIZATION

This is an abstraction operation whereby instances of entities (that share a common property or properties) are viewed as instances of a single, higher level entity. For example, we may generalize the entities aircraft, land vehicle and boat into the more generic entity vehicle. Generalization thus establishes the 'is-a' relationship between instances of an entity and a more generic entity, so that an instance of the lower level entity 'is-a' instance of the higher level entity (but not vice versa). The use of generalization to form subsets and supersets has been discussed above.

An important feature possessed by entities in a generalization hierarchy, from the point of view of reducing detail on the entity model, is that as a lower level instance 'is-a' higher level instance it accordingly 'inherits' all the properties (entity and attribute relationships) of the higher level entity. This is termed the 'inheritance' principle. For example, in Fig. 8.18, a cost accountant inherits properties of accountant, which inherits properties of employee.

AGGREGATION

Aggregation is an operation whereby objects are aggregated to be 'part-of' a higher level object. For example, all the objects in Fig. 8.18 might be aggregrated to form a 'production' entity, which may be related to other entities on a similar level (not shown) such as R & D, product quality and building maintenance. The complement operation, 'de-aggregation', is frequently used when developing high-level entity models to produce a set of hierarchically related models. An overview entity model may also be produced by aggregating all attributes into their entities, so only entities and relationships are shown.

Relational data model

The relational data model is an approach to modelling data based on the mathematical theory of relations, and is used in the analysis phase to model objects in a 'bottom-up' manner. This means that it is used to produce what are termed *relations* from input and output forms of an existing system, such as invoices, reports and transaction input. The relations are then transformed into entities, attributes and relationships.

The procedure may also be used as a bottom-up check that an existing entity model is a correct model of the data.

RELATIONS

We may explain the notion of relation in terms of its correspondence to the components of the entity model. An entity in the entity model corresponds to a relation or, to use the term for its diagrammatic representation, a table. The entity name becomes the table name, the entity attributes become the table columns and the attribute names become the column names.

We may visualize entity instances as rows in these tables, and, for any given entity instance, the values in the columns for the row are the attribute values of that entity instance. The entity instance itself is represented by an identifier, termed the primary key. Figure 8.19 shows the product relation, in table form, with two instances. Each row is termed a tuple.

It is common to give examples of relations by showing such a table containing actual data values in the rows (termed the extension of the relation). However, in a specification, we only need to assert the existence of the relation and its attributes, so we use a notation (termed the intension of the relation) that shows the relation name, together with its attributes. For

Product

Product-type	Product-name	Product-number	Price 1	Price 2		Price n
0	Soprano bell	Sop92/1456	567.00	525.00		475.00
1	Medium clapper	Mcl91/325	45.00	40.00		34.00

Figure 8.19   A part of the product relation

The convention we adopt is to specify such rules on the entity, that is on Row 2.

We will not discuss all the different types of rule in Table 8.1, but we will present three types of rule on row 1, that is those rules that involve attributes. In addition, we shall discuss what is often known as relationship optionality, and show how it is modelled as a rule.

Attribute cardinality rule (1.1)

This rule restricts the number of values of a given attribute. For example, we may restrict the number of values of the material-code attribute to 300. A typical expression would be, for example:

CARD(MATERIAL-CODE) $\leqslant$ 300

Attribute set rule (1.2)

Here we want to restrict the actual values that an attribute may possess. For example, to restrict the values of the price attribute to between 50 and 2000, we can write:

PRICE IN {50 . . . 2000}

To restrict the values of the product-number attribute to between 1 and 99 999:

PRODUCT-NUMBER IN {1 . . . 99 999}

Attribute function rule (1.3)

The rule that restricts attribute values to be derived from other values may be expressed as a function. For example, if danger-level values are to be derived from the reorder-level attribute:

DANGER-LEVEL = REORDER-LEVEL * 80%

Optionality rule (3.2.1)

This concerns the participation of set instances in a relationship. Either *all* instances must participate in the relationship or *not all*. For example, from Fig. 8.18, there may be a rule that all instances of product must participate in the relationship product–clapper. In user terms, this means that all products must be associated to a clapper. We express this rule as:

PRODUCT : CLAPPER = PRODUCT

If not all instances have to participate, as for the container–container relationship, we do not need to assert this rule for the relationship.

Rules are specified declaratively and are only expressed in terms of objects in the structural component. Rules are kept separately in a rule model and are easily related to their objects and to the processes that require them, as objects are part of the rule expression. One copy only of rules removes the problem of duplication of rules over processes.

Rule specification can help as a check on the contents of the structural component. For example, a rule referring to an object not modelled might indicate either that the object should be modelled or that the rule is outside the system.

Principles of analysis

This section will present some principles that are useful to bear in mind when carrying out analysis to help achieve the aims of a correct and complete specification:

1. To model *completely* — to model in all the detail required on the conceptual level of abstraction.
2. To be *precise* — to expose any ambiguities, redundancies or inconsistencies in the requirements (often the case where multiple users exist). A precise language helps us to achieve this.
3. To be *abstract* — to avoid any detail regarding design or implementation. Such detail is firmly relegated to lower levels of the development process. User terms and concepts must be retained so that a description based on the specification would be recognizable to a user.
4. To be *declarative* — to avoid procedural forms as much as possible. It is not clear to what extent there is a trade-off between the formality and the declarative nature of a language required for the declarative specification of processes. However, we should postpone, to the latest possible stage, any descriptions of processes that resemble program code, either by their level of detail or their level of abstraction.
5. To avoid *redundancy* — to model facts in the requirements only once.
6. To be *natural* — to aim for a one-to-one correspondence with real-world objects. This means that if the user is accustomed to regarding certain components of the application as single concepts, for example a static object, or a single process, the modelling language should preserve this single nature by providing constructs (language elements) that model these components. We may contrast this approach with one where an application object may be modelled in many modelling language constructs, or vice versa.

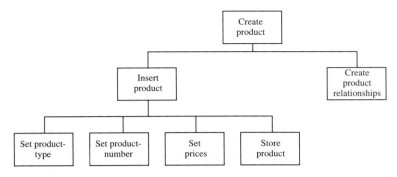

Figure 9.2 Refinement of create product process

Data flow diagram

The process decomposition has the advantage of showing the refinement of one process in great detail. However, we often wish to examine the interrelated inputs and outputs of processes, and a widely used method that shows such process relationships is the DFD, or *data flow diagram* (Gane and Sarson, 1979; de Marco, 1979).

Such a view gives a wide overview of a system, and is particularly suited to those systems where data is transformed by a series of processes, the sequence forming a logical whole, such as the transaction processing cycle discussed in Chapter 4.

DATA FLOW DIAGRAM SYMBOLS AND RULES

Data flow diagrams are made up of five symbols, which represent important system components. Symbols vary, and Fig. 9.3 shows a set of symbols in common use.

Processes are shown by circles and each *process* has a unique name and number. The number may indicate the level of the process (for example 2, 2.1, 2.1.1 and so on). All processes must have an input and an output, consisting of a *data flow*, which is shown by a line joining the process to other symbols on the diagram. The direction of the flow is shown by the arrow and the line is labelled with the name of the data flow. *Events*, which initiate

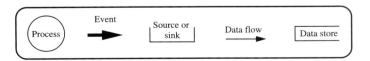

Figure 9.3 Data flow diagram symbols

processes and which take place outside the system, are shown by thick arrows, labelled with the name of the event, and the event is placed by the process which it initiates.

External entities, shown by an open square, are outside the system in the sense that the system designer does not specify their behaviour or does not record information about them. Typically, they supply system input and receive system output. Input entities are often termed *sources*, while those receiving output are termed *sinks*.

Finally, data may be stored in a *data store*, shown by an open rectangle. This acts as a buffer, as there may be a time delay between processes where, for example, process B may not necessarily start immediately after process A finishes. Each data store has a unique name, and processes may output data into a data store and input data from a store. Data cannot flow directly between stores.

DFD LIMITATIONS

The data flow diagram originates from structured systems analysis in the 1970s, and is not entirely suitable as an analysis tool as it models *data* flow and uses data stores. Although this emphasis on data may correspond closely to service-oriented applications, such as banking and insurance, it does not apply so well to manufacturing applications, such as the Eurobells △ production system, which is partly concerned with the flow of goods and products.

However, the data flow diagram may be used, somewhat awkwardly, by substituting the notion of product for data, so that the diagram models product flow and store as well as data. In addition, if an entity model has been developed, items referenced on the DFD should be components of the entity model. The DFD must be kept on a conceptual level, only concerned with *what* products are used and stored, not how or where.

EXAMPLE DFD

Figure 9.4 shows a DFD that models the Eurobells △ production system, where the processes are the level 1 processes from Fig. 9.1.

A warehouse is a source, which sends raw bronze to the organization. This is input to the receive raw bronze process, which adds the bronze to the stock of raw material. The next stage is where the melt and mix process begins, which takes raw material as input, producing mixed bronze, which is input to the cast bell process, which produces a bell.

The tune process takes a bell as input and produces a tuned bell. The assemble product process produces a product from a bell and clapper, also generating product details for production control, which constitutes a sink. The package product process packages a product in a container and the ship process finally takes a product and sends it to a distribution organization, along with a shipping note for sending to the customer.

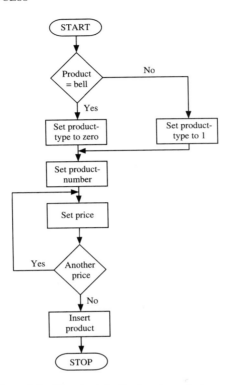

Figure 9.7   Flowchart for the insert product process

```
BEGIN
    Receive product transaction.
    IF product = 'bell'
    THEN
        BEGIN
            Set PRODUCT-TYPE to ZERO.
        END
    ELSE
        BEGIN
            Set PRODUCT-TYPE to 1.
        END
    Set PRODUCT-NUMBER.
    WHILE there are more prices DO
        BEGIN
            Set next PRICE.
        END
    Insert product
END
```

Figure 9.8   Structured English for insert product process

statement, while 'pseudocode' uses programming-like constructs such as
DO-WHILE and END-DO statements.

Types of structured English have a disadvantage in that they become hard
to follow when there are many decisions and consequential actions. The
proliferation of BEGIN and DO makes the text longer and harder to read.
Their main use is as a logical basis for programming.

Action diagram

The *action diagram* uses a mixture of narrative and graphical notations to
show process structure. It is described in Martin and McClure (1989) and
Gane (1990), and is a simplified version of the more elaborate Warnier–Orr
diagrams (Orr, 1981; Connor, 1985). Although they have been used mainly in
the United States, they are included in some recent CASE tools available in
the United Kingdom (see Chapter 10) and their use is expected to grow. The
structured English example above is shown in action diagram form in Fig. 9.9.

There are two diagrammatic aspects to the action diagram. Firstly, a
bracket is used to group a set of related actions which constitute a control
structure and, secondly, bracketed actions may be nested within other
brackets using indentation. The main usefulness of action diagrams is thus to
show, diagrammatically, hierarchical relationships within a process.

Figure 9.9 shows syntax and symbols from the action diagrams used in the
Information Engineering Workbench (IEW) CASE tool. Iteration is shown
by a thicker bracket and a double line at the top of the bracket. The dash
attached to the ELSE statement indicates that control may flow to processes
within the scope of the ELSE or the scope of the IF, but not both.

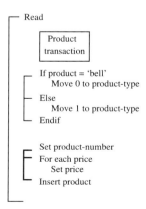

Figure 9.9   Action diagram for insert product process

As for structured English, action diagrams are used to show the lowest level of process detail before translation (manually or automatically) to programming code or 4GL statements. The diagrammatic symbols make the diagrams more complex than structured English, and thus harder to change, so this type may be best suited to use within a CASE tool.

Nassi–Schneiderman chart

Like action diagrams, *Nassi–Schneiderman charts* are structured diagrams (Nassi and Schneiderman, 1973) that show *process control structure*, and are again used mainly in the United States. Figure 9.10 shows the three symbols used and Fig. 9.11 shows how the symbols are combined for our example. This chart is meant to be used with a top-down design approach. Major loops or actions are drawn first, and then minor ones. They are similar to action diagrams as they use indentation for sub-processes, but this only extends to iteration, and not selection.

Although it may be argued that selection has a symbol of its own, and is therefore clear, it is not possible to show subsets of related actions as clearly as action diagrams. It is also likely that action diagrams are easier to change,

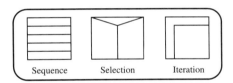

| Sequence | Selection | Iteration |

Figure 9.10   Diagrammatic forms for control structures on Nassi–Schneiderman chart

Do for product transaction	
Case of product-name	
Product = 'bell'	Product = other
Set product-type to zero	Set product-type to 1
Set product-number	
Do for each price	
Set price	
Insert product	

Figure 9.11   Nassi–Schneiderman chart for insert product process

as they have only the bracket symbol. The selection symbol of the Nassi–Schneiderman chart may vary in size depending on the number of selections.

As for action diagrams, these are used for detailed process levels and may be more suited for use in a CASE tool environment.

Decision tree

Decision trees (de Marco, 1979; Avison and Fitzgerald, 1988) are used to describe a process (or process fragment) that contains a number of alternative actions to be selected. A tree consists of a set of nodes, which are conditions, and a set of unlabelled edges, while the leaves (rightmost nodes) contain the actions. The first (leftmost) node shows the name of the process. Trees are usually shown left to right, although they may be shown top to bottom.

A decision tree for the select container sub-process of the package product process from Fig. 9.1 is shown in Fig. 9.12. Within the process, different types and qualities of bell are determined for matching with different sizes and finish of container.

Tree styles are not standardized, and an alternative for Fig. 9.12 is to use a predicate form, where edges are labelled only true or false, as in Fig. 9.13.

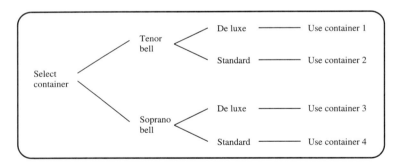

Figure 9.12   Decision tree for the select container process

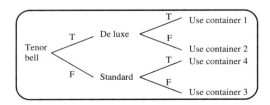

Figure 9.13   Predicate form of decision tree

This has fewer nodes, but is not so informative, as only a tenor bell is mentioned, compared to Fig. 9.12, which refers to both soprano and tenor bells.

The tree is normally used for complicated conditions and dependent actions within processes and may be suitable for working through with users. The tree is used in preference to the direction table (see below) as it shows the likely sequence of events dependent on conditions. It is used where there are a *large number of conditions* that determine the appropriate action.

Decision table

Figure 9.14 shows a *decision table* (de Marco, 1979; Avison and Fitzgerald, 1988), containing actions in what is termed the 'action stub' in the bottom left-hand part of the table, with the conditions in the 'condition stub' in the top left-hand part. A condition is expressed in predicate form, that is it expects an answer 'true' or 'false'. All possible combinations of true and false responses are recorded in the top right-hand part of the table, termed the condition entry. In the corresponding columns of the bottom right-hand part, the action entry contains an ×, indicating the appropriate action to be taken.

The decision table is more suitable for large numbers of *actions* than the decision tree, and it is also easier to check that all conditions have been catered for. Where there are many conditions, it is inferior to the decision tree, as, for n conditions, the table requires 2^n condition entries. Four separate conditions would therefore create a table with 16 columns in the condition entry. In general, its use is similar to that of the decision tree.

Entity life history

An *entity life history* (ELH) (Jackson, 1983; Downs, Clare and Coe, 1988) shows the *processes*, together with their *control structure*, that take place on

Condition stub		Condition entry		
Tenor bell	T	T	F	F
De luxe model	T	F	T	F
Use container 1	×			
Use container 2		×		
Use container 3			×	
Use container 4				×
Action stub		**Action entry**		

Figure 9.14 Decision table for select container process

one entity only. It is a diagrammatic way of specifying processing which is on a level of detail midway between that of the process decomposition and structured English. Only *update processes* are specified on the ELH.

All objects referenced by the process should be components of the entity model which are either an entity or are related to the entity, such as its attributes or relationships. An important point about the ELH, often overlooked, is that the processes take place on a *particular instance* of the entity, and not any instance.

Figure 9.15 shows the ELH for the bell entity. The entity is the root node, with the other nodes being the processes and their refinements that operate on the bell entity.

- *Sequence.* The ELH is often constructed by considering the processes that constitute the 'birth' and 'death' of an entity, and then the processes that constitute its 'life' (within the system under consideration). Figure 9.15 shows the allowed sequence of process execution on the bell entity diagrammatically from left to right. Thus, a given bell instance must have the cast bell (birth), tune (life) and assemble product (death) processes executed only in that sequence. If processes took place on any instance, as opposed to a particular instance, then the sequence would not apply, as bell B might be tuned before bell A is cast.
- *Iteration.* This is shown with an asterisk, denoting a process that can execute zero or many times. Figure 9.15 shows that a bell instance may be tuned zero or many times, but cast and assembled into a product only once.
- *Selection.* There is a choice of processes to execute within tuning, and the circle in the top right-hand corner of the process box indicates that only

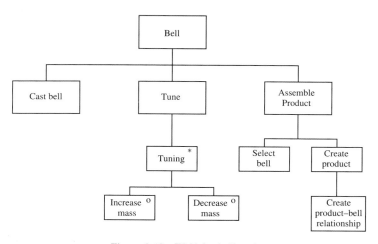

Figure 9.15 ELH for bell entity

one of the processes can be executed. The tuning process consists either of increase mass or decrease mass, but not both. The diagram does not show the conditions under which a selection branch is taken, that is, the condition that must be true to increase as opposed to decrease mass.

There are two main uses of the ELH. Firstly, it acts as a check on the objects used by processes. This might uncover some objects that have been omitted from the entity model, or there might be some objects that are never referenced by processes. Secondly, as stated above, it may be used to specify process control structure on an intermediate level of detail.

It may be useful on an overview level, but only where correct process sequence is central to the application. This occurs only when a particular object is the most important, so that the ELH of that object spans the major processes. In Fig. 9.15, this is not the case, as the bell object is only involved in two level 1 processes. Again, in Fig. 9.15, only a small part of the assemble product process concerns the bell object.

A disadvantage is that it is often necessary to introduce redundant boxes (for example, tuning in Fig. 9.15) to follow diagram syntax rules.

Summary of process phase products

Table 9.1 summarizes the above discussion with respect to the different means for specifying processes. It can be seen that a detailed process specification may be difficult to validate with users, apart from decision trees and tables.

Table 9.1 Characteristics of different types of process specification

	Form of expression[†]	Level of detail[‡]	Purpose[§]	Components[¶]	Suitable for users[††]	Easy to modify[‡‡]
Process decomposition	D	O/M/D	G	P, E	Y	Y
Data flow diagram	D	O/M	G	P, E, D	Y/N	Y
State transition diagram	D	O	H	P, C	Y	N
Flowchart	D	D	G	P, C	N	N
Structured English	T	D	G	P, C, D	N	Y/N
Action diagram	T/D	D	G	P, C, D	N	N
Nassi–Schneiderman chart	T/D	D	G	P, C, D	N	N
Decision tree	D	D	S	P, C	Y	N
Decision table	D	D	S	P, C	Y	N
Entity life history	D	M	U	P, C	Y/N	Y

[†] Diagram (D), text (T).
[‡] Overview (O), medium (M), detailed (D).
[§] General (G), HCI (H), update (U), selection (S).
[¶] Process (P), control structure (C), data (D), event (E).
[††] Yes (Y), moderately suitable (Y/N), not suitable (N).
[‡‡] Yes (Y), no (N).

Object-oriented analysis

CURRENT APPROACH

The object-oriented approach to analysis (OOA) is fairly recent and is very different from the more traditional approaches described above. We will only present the main points here, but further discussion may be found in Shlaer and Mellor (1988), Coad and Yourdon (1990) and Korson and McGregor (1990).

Under an object-oriented approach, a specification consists of only one basic type of component — the *object*. Objects have different types of relationships with each other, and the approach specifies objects and then relationships to build a network for eventual mapping to design and on to an implemented system.

A fundamental object-oriented principle is that an object and the processes involving that object (termed *services*) are always considered together as a whole. In addition, attributes of an object are always considered together with their object. This principle, which is a type of abstraction, is termed *encapsulation*.

In addition to being an abstraction of services and attributes, an object may also be an abstraction of different kinds with respect to other objects. The two most common abstractions are generalization (for example, a vehicle object is a generalization of ship, car and plane objects) and aggregation (for example, a car object is an aggregation of engine, chassis and wheels objects). The generalization abstraction is important as it allows *inheritance*, where objects may inherit attributes and services of higher level objects, while specialized objects may have their own objects and services.

Processing is viewed from the perspective that services communicate via *messages*. A service of a sending object sends a message to a service of a receiving object and may receive a response. Services send messages (which may consist of data or an instruction to initiate processing) to get other services to do processing for them. This is similar to the familiar idea of DFD processes sending data and transferring control to one another. A service may be specified in, for example, structured English and will refer to any messages to be received from other services, as well as messages to be sent to services.

Various mechanisms exist for grouping services belonging to different objects into higher level processes that correspond to a transaction. For example, the HOOD (hierarchical object-oriented design) method (HOOD, 1989) uses the concept of parent and child objects to group services into higher level processes.

Advantages

1. A specification consisting of object-oriented objects can be agreed upon more readily by all in the development process, as there is often a large

measure of agreement about what the basic entities are. In contrast, a problem of process decomposition is that different individuals usually produce different decompositions, making agreement over a correct or complete specification difficult.

Given the agreement on basic components, it is possible to engineer specifications by assembling tried and tested components (that is, objects) into a final specification. Of course, a library of such components takes time to build in an organization.

2. The object-oriented approach to analysis is claimed to reduce maintenance costs as it is more stable over time and hence is easier to change.
3. The object-oriented approach to analysis maps down to the design phase (which is becoming object oriented) more easily, reducing the problem of traceability between different phases.

Disadvantages

1. Although objects are easier to agree upon than functional decompositions, services of objects may be specified on different levels of detail, or detail may reside in the service of a receiving object as opposed to a sending object. The 'insides' of objects may thus be very different, although similar on the surface, and so components may not be so easily reusable or agreed upon.
2. It is not clear how rules are integrated with services.
3. It is not clear how common actions (actions that take place on more than one object) should be handled.

IMPROVED APPROACH

Steps

An object-oriented approach to analysis will now be described in outline, which is based on objects (albeit of a different kind) but which incorporates rules and takes account of transactions and events. It will be given in step form:

Step 1. Build system overview.
    (a) Build entity model
    (b) Build rule model
    (c) Build process model
Step 2. On the process model, identify events and corresponding processes constituting transactions.
Step 3. Identify a transaction and, from the entity model, generate relevant primitive processes for the object(s) referenced for each transaction process. For example, if we choose the assemble product transaction and, within this, the create product process, we may use the entity

model in Chapter 8 to *generate* the insert primitive processes for the product entity, as shown in Fig. 9.16. All attributes and entity relationships are known from the entity model.

Step 4. The lowest level operations may be modified to suit the process. In this example, insert requires all the attributes, but others (for example, change) may require only some.

Step 5. Incorporate rules from the rule model — generate the create product object. This final step generates a more detailed version of Fig. 9.16, and this object is shown in Fig. 9.17.

Generated object

An *object* is an application-oriented processing unit, related to an entity, which is intended to preserve the integrity of the entity, that is to make sure that no rules are violated by any actions on the entity. An object consists of several components: pre-conditions, actions and the chief component, which

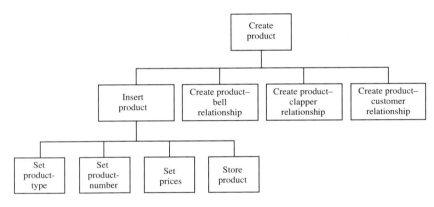

Figure 9.16 Generated insert primitive processes for create product

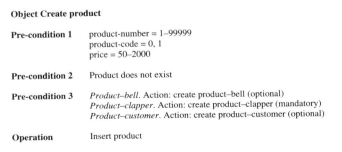

Figure 9.17 Generated create product object

we shall term the operation. It will be seen that almost all the processing in this object is a preliminary to actually creating an instance of product.

Before the operation can execute, different types of pre-conditions must be met. A pre-condition is something that must be true for the operation to execute. In addition, certain actions (such as creating relationships) may also be necessary, which also depend on pre-conditions. The rule model is the source for pre-conditions:

Pre-condition 1. Attribute values of product are checked here. For example, the rule model in Chapter 8 may contain rules that product-code must be 0 or 1, product-number is the range of 1 to 99 999 and prices should be in the range of £50 to £2000. Statements will be generated for the detail of rule checking.

Pre-condition 2. It should be true that this product instance does not already exist. This pre-condition is automatically generated for any insert.

Pre-condition 3. The entity model shows that product may have three relationships, with bell, clapper and customer. Statements may be generated to create these relationships, and if our rule model contains an optionality rule for one of these relationships, product and clapper, this relationship must be created, and is therefore marked mandatory.

Dependent actions, if these pre-conditions are met, will create relationships with the instance of product.

Operation. If all pre-conditions are met (including the successful creation of relationships), the instance of product is inserted.

The important difference between our type of object in Fig. 9.17 and the type of object suggested by the conventional object-oriented approach above is that our object is oriented to only one type of primitive process. This enables it to be used, as a whole, in higher level processes. For the conventional object-oriented object, containing all the processing on an entity, parts of the object are used in different higher-level processes. This makes the specification more complex, as only certain parts of objects must be referenced by other objects.

Furthermore, our object is a unit of integrity, incorporating all relevant rules from the rule model. In addition, it can be generated completely from the (declarative) entity and rule models, minimizing error and maximizing consistency between different parts of the specification. This integrative feature also avoids the problem of separate entity and process development. Detailed error handling may be generated or inserted later on in design in the pre-conditions.

Problems in analysis

Problems with analysing and specifying processes are similar to those concerned with objects and rules, discussed at the end of the previous chapter. It is not clear whether the object-oriented approach for process specification will bring about any reduction in the errors made by analysts, although it brings together object, rule and process in the specification, making prototyping the specification easier. Demonstrating the effects that processes have on objects is a good way of presenting processes to users for checking (Warhurst and Flynn, 1990).

Summary

In this chapter, we have discussed various means for representing processes. Process decompositions and data flow diagrams are the most common forms for use on the overview level and are complementary, as the data flow diagram basically adds data flow to the process decomposition. Events may be shown on both types of diagram and are important, as they identify groups of related processes which will form transactions. The state transition diagram, emphasizing states and the sequence in which they are reached, is most suitable for specifying HCI processes. Flowcharts, action diagrams and Nassi–Schneiderman diagrams are all used for specifying low-level detail diagrammatically, while structured English is a textual form. Decision trees and tables are used for specifying complex selection conditions which are process fragments.

In the last section, we described the object-oriented approach to specification that is likely to dominate in the future. However, there are problems with integrating rules and services.

One way of solving this was described using a hybrid approach consisting of top-down analysis, employing traditional means such as process decomposition and DFD for transaction analysis and then composing the transactions in a bottom-up manner using objects. Our type of object is different to the normal object-oriented object. Furthermore, improvements are made by *generating* the objects from the entity model and the rule model. This reduces designer intuition, reducing error, and ensures specification consistency.

Discussion questions

1. List the main components of a process specification.
2. Do you think that a fourth type of process control structure should exist, modelling the parallel execution of processes? In what diagram would you incorporate this and how would you show it?
3. Why do you think the DFD, ELH and structured English are unsuitable for showing to users?

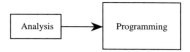

Figure 10.1   Pre-method model of systems development process

ties within these phases were not defined, varying between individuals, and had no start or end and no defined products existed.

The level of abstraction implicitly emphasized was the programming level, as program execution efficiency, in terms of either execution speed or main memory size, was the overriding consideration. Types of assistance typically available consisted either of efficient algorithms for common programming problems, such as master-detail batch tape update, or diagrammatic techniques such as the program flowchart, which was used to help code the more difficult parts of programs.

Work began with analysis and continued until the system was in operation. The size of the boxes in Fig. 10.1 reflect the approximate amount of resources expended on the areas, and the tendency was (and still is) to concentrate resources on programming at the expense of analysis and design, especially if there was pressure for the system to be installed, as code was the only visible product. Documentation had a low priority.

An example of a method of this type, which had a commercial following in the United Kingdom, was that recommended by the National Computing Centre (NCC). The method, which evolved in the 1960s, is described in several publications concerned with the analysis and design of data processing systems (Daniels and Yeates, 1969; Lee, 1979). Although this method contained different phases (analysis, design, implementation and maintenance and review) they contained few defined activities or techniques and had little or no interdependence. They consisted mainly of advice, such as guidelines for forms design for input and output documents.

STRUCTURED METHODS — 1970s

Structured methods mark a significant point in method evolution. They are characterized by the gradual progression from the simple model of systems development to one that increasingly differentiated between the development activities. The problems addressed were discussed at a well-known NATO conference in Garmisch, Germany, where the term software engineering was first used (Naur and Randell, 1968).

Structured programming

- *Programming.* The central principle of structured programming was that only three constructs, sequence, selection and iteration, were sufficient to code programs with one entry and exit point. The principle was the result of the work of two Italian mathematicians, Bohm and Jacopini (Bohm and Jacopini, 1966), and that of Dijkstra, who became famous with his 'Go To statement considered harmful' letter (Dijkstra, 1968). Coding could thus be reduced to the use of these three constructs, and the proliferation of undisciplined 'spaghetti' code reduced.

The program flowchart was given a structured companion by the Nassi–Schneiderman chart (Nassi and Schneiderman, 1973), which was a type of structured flowchart, only allowing diagrammatic representations of the three structured constructs.

- *Project management.* A project that took place at the *New York Times* (Baker, 1972) introduced elements of project management into the development process. In addition to using structured programming as a project standard, as well as the top-down program structuring approach, a program production library was set up, organized by a program librarian, to centralize documentation and to keep track of programs and program changes. The programmers were also organized according to a concept termed the chief programmer team.

An activity that distinguished testing from programming, termed code inspection (Fagan, 1976), defined desk-checking aspects of testing by instituting a walkthrough procedure, with defined activities and individuals, to check code before it was tested by execution.

Structured design

The design of programs and files had largely been left to individual intuition until now. General program design techniques began to appear that had the effect of differentiating physical design activities from those of programming. There were two main approaches, the functional and the data structured approach.

- *Functional approach.* This approach, developed in the United States (Stevens, Myers and Constantine, 1974), produced a design for a system or program by decomposition into a number of hierarchical functions or modules, expressed these on a structure chart and then applied design rules, using the concepts of coupling and cohesion to determine the quality of the design. The rules were later elaborated in Myers (1975) and in Yourdon and Constantine (1979). Data flow diagrams were used to identify major data transformations and sub-systems and to act as a basis for the production of structure charts.
- *Data structured approach.* This was presented in the work of Jackson (Jackson, 1975) in the United Kingdom and Warnier (1974) in France. Under the Jackson Structured Programming (JSP) approach, the structure of input and output data, shown diagrammatically in structure charts, determined the structure of the program, that is the different hierarchical

functions the program should have. The work of Warnier was made more generally available by the development of Warnier–Orr diagrams (Orr, 1977), which used a bracketed type of indented narrative to show function hierarchy similar to action diagrams. Data flow diagrams were not used in this approach.

Structured analysis

The basic principles of structured systems analysis were set forth in several influential publications from the late 1970s (Ross and Schoman, 1977; Gane and Sarson, 1979; de Marco, 1979), and grew from the use of the data flow diagram (DFD) in structured design.

- *Analysis.* The emphasis was on providing a non-technical picture of the user's requirements and the medium advocated was the DFD, which was used to understand and model the application in terms of the hierarchical decomposition of organizational processes and data flows (although the data flows were usually a secondary consideration), rather than the programs and files of physical design.

  Abstraction away from physical considerations was thus occurring, as the top-level DFD gave an overview of the system, which was considered to assist in the clarification of requirements. Also provided were facilities for more detailed specification, such as minispecifications, data dictionary, decision tables and trees, and others.
- *Walkthroughs.* Products were more clearly described and walkthroughs setting out checking procedures for the products were also introduced (Yourdon, 1978). Validation with users was thus being introduced into the development process.

  The JSD method (Jackson, 1983) was a later UK method, which also modelled organizational processes and data but which was based on the entity life history approach and did not use hierarchical decomposition.

Summary of structured methods

- *Process oriented.* Structured methods were process oriented, in that the methods and techniques they provided emphasized the analysis and design of processes, as opposed to data. The level of abstraction of processes had been raised from programs to high-level data transformations, but data was still viewed in physical terms.
- *Normalization.* There was a complement to this process orientation in the technique of normalization, described in Chapter 8. This came into use as a data and file design technique in the late 1970s and was not initially integrated into the structured approach. The input to normalization was generally a set of data files or input/output documents which were

normalized into a set of relations, suitable for use in a relational database (Codd, 1970). More recent publications on structured analysis now attempt to integrate normalization and process design.

- *Model.* By the end of the 1970s, the more widespread use of structured methods, involving the emergence of distinct phases and defined phase products, was reflected in the adoption of a process often termed the 'waterfall' (Royce, 1970) or 'avalanche' model, as shown in Fig. 10.2, so called as it tended to create an avalanche of paper documentation as the project progressed. The appearance of maintenance on the model was due to realization of its increasing importance, as its share of the software budget was rising steadily.

The more precise definition of phase products also allowed the development of walkthrough activities for checking.

DATA-ORIENTED METHODS — 1980s

Deficiencies of structured methods

However, dissatisfaction grew with structured methods, as many errors were still found at the test phase or in operation and maintenance. A study of the main structured analysis methods reveals that, in general, they assumed that users had a current (usually manual) system, which they wanted to computerize for increased efficiency. The user requirement was thus assumed to be fixed, implicitly on the functional organizational level, and improvements would result simply from making the old system more efficient, without the need for a new or extensively modified system.

However, as new systems were required that had not existed before, such as management and strategic information systems, which required significant user participation for establishing requirements, the structured analysis approach ran into difficulties. Methods that were more successful at establishing this new type of user requirement, as well as using terms that users could understand, came to be needed. The analysis phase in Fig. 10.2 had been expressed on a relatively low level and much of it came to be known later as logical design.

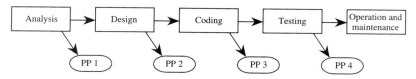

Figure 10.2 Avalanche model of systems development showing the production of phase products (PP)

Case Study 4 — the university library

The case study adopted is small, given the chapter space available, but it is part of a realistic case study used in commercial courses for method training. The user requirement may be seen in Fig. 10.5 and concerns part of the activities of a library.

Information Engineering

PHASES AND PRODUCTS

Figure 10.6 shows that Information Engineering (IE), following the description given in Finkelstein (1989), consists of three main phases, analysis, design and generation. It also contains a strategic planning phase that is not relevant here, which is discussed in Chapter 14. The main products occur in the analysis and are different levels of data models, termed strategic, tactical and operational data models.

ANALYSIS PHASE

The analysis phase consists of four stages: project scope, strategic modeling, tactical modeling and operations modeling.

Project scope stage

This first stage is a preliminary stage whose purpose is to scope the project. Its steps are:

- *Identify project area.* The project boundaries are set, identifying the areas to be included and excluded. This prevents the project expanding

> The aim of the library is to serve its members, and other libraries, by providing high quality book services at a fair cost. A member may join and leave the library. Before leaving, the member may loan and renew books. A book reservation may be made (either by title or for a specific book volume) and cancelled. After a member leaves, tidying-up will return any books and cancel any outstanding reservations. Books are obtained either by acquisition or by being borrowed from other libraries (swapped-in), and disposed of by being given to a local charity or swapped-out to their original library. A book may be sent for binding, and received back, while in the library. An automated information system is required to improve efficiency.

Figure 10.5 Library case study

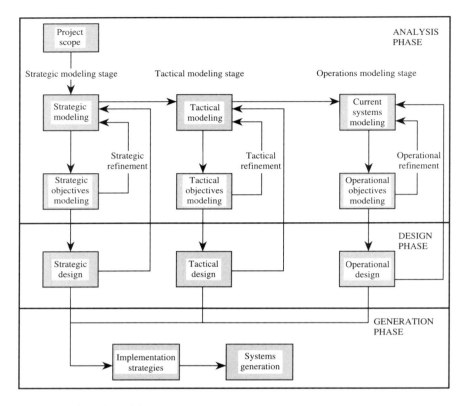

Figure 10.6 Overview of three main phases of Information Engineering (adapted from *An Introduction to Information Engineering*, C Finkelstein, Addison-Wesley, 1989)

uncontrollably to other, irrelevant areas, which can jeopardize the success of the project. Organizational priorities relevant to the project must be set, which may be defined from the strategic statements obtained from strategic planning or from the management questionnaire (see Chapter 14). Project aims, deliverables and completion time are set.

- *Select software tools.* The tools to be used for the strategic and tactical modeling stages are selected. IE suggests that these include a data dictionary, a diagramming tool and an integrated word-processing system.
- *Establish initial project plan.* Project planning is based on estimates of the numbers of project areas that are involved. Each project or strategic area typically contains 50 to 90 strategic entities and is refined into tactical areas of 30 to 40 entities each. Estimates of the time needed for each of the stages and phases are then made.

- *Establish project teams.* Of the project teams 80 per cent should be composed of managers and those users with the greatest knowledge of the project areas. Remaining members may be analysts or data administrators. Team size is suggested as six members.
- *Set project budget and funding.* Based on project size, number and size of teams, hardware and software support, the project budget is set.
- *Schedule Information Engineering workshops.* As modelling is mostly done by users, these five-day workshops train the users in IE methods and techniques.

Strategic modeling stage

This stage produces the *strategic model*, which contains entities mainly of interest to senior management.

- *Strategic modeling.* The initial input to this phase is the set of relevant strategic statements from strategic planning (see Chapter 14). These consist of the mission, purpose, concerns, issues, goals, objectives, policies, strategies and tactics of the project area. These are analysed and used as catalysts to identify strategic organizational data.

  Heuristics are given to assist in mapping organizational features on to data. For example, policies and issues map to entities, goals and objectives to attributes, and strategies and tactics to associations.

  The output, the strategic model, consists of high-level strategic entities, and is produced by applying *business normalization*, which is similar to the normalization techniques discussed in Chapter 8. The model is expressed in a *data map*, which is similar to an entity diagram, or as an *entity list*.
- *Strategic objectives modeling.* This phase reviews and identifies criteria for performance monitoring using goals, objectives, policies, concerns and issues, and defines strategic data required to measure performance. Performance ranges and controls for early warning systems may be defined. Strategic attributes are added to the strategic model by this phase.
- *Strategic refinement.* This is an iterative step which uses business normalization to refine the strategic model, with the aim of identifying 'hidden' entities. The strategic statements referred to earlier may also change here. Standard terminology and performance rules are identified.

The final strategic model is used to identify *strategic sub-models*, which will form the basis for the tactical data models in the tactical modeling stage.

Tactical modeling stage

This stage refines the strategic model by producing different *tactical data models* for *key functional areas*, of interest to middle management, which are based upon the products, services, markets and distribution channels with which the organization implements the basic strategy.

- *Tactical modeling.* The relevant data is identified in detail for each functional area. This may include data used to derive strategic data. A tactical model is produced for each area.
- *Tactical objectives modeling.* The data required to measure the achievement of tactical objectives on this level is identified. Exception report requirements are noted as well as criteria for decision making by middle management.
- *Tactical refinement.* Normalization is applied to the tactical models and, as before, hidden entities may be identified. Models across different functional areas may be compared and related data may be grouped together.

Operations modeling stage

This stage is concerned with the operational, day-to-day level of the organization. It firstly looks at current manual or automated systems and identifies the data used, as well as any interfaces that will be required. Each tactical area is then refined for its operational data.

- *Current systems modeling.* This cross-checks strategic and tactical data against the data currently used by the organization. Documentation, enquiries and computer files may need to be consulted. Data that was overlooked may be discovered.
- *Operational objectives modeling.* Data required for the day-to-day measurement of objectives are identified, as well as exception reports.
- *Operational refinement.* Operational data are obtained by refining a tactical area and *operational data models* are produced.

DESIGN PHASE

The design phase is built around the *design dictionary*, which is the central store (intended to be automated) for the data identified during the analysis phase. The data entered into the dictionary consists of detailed descriptions for each attribute, entity and association, including a description of their purpose or use within the organization, and the data types and lengths of attributes. Cross-references between elements are also made. A *strategic planning dictionary* is also maintained to record mission, objectives and so on, cross-referenced to design dictionary elements. There is tool support for interface design (see below).

Generation phase

With automated support, database definition statements for the application group defined above may be generated.

Structured systems analysis and design

INTRODUCTION

There are a number of variants of what are often referred to generically as *structured* techniques. We shall refer to the common points as the *method*, drawing attention to important distinctions. The main sources for structured analysis are Gane and Sarson (1979) and de Marco (1979), and for structured design, Yourdon and Constantine (1979) and Myers (1975). The method does not consist of a prescribed set of phases, but is more a set of techniques to be used as the designer sees fit, although authors do make suggestions for some activities.

The aim of structured analysis is to produce a logical description of the data and procedures of the required system. Structured design is defined as 'the art of designing the components of a system and the interrelationships between these components in the best possible way' (Yourdon and Constantine, 1979). This is a rather high-level definition, and we can say that structured design aims to produce a physical design of an implementable system from a structured analysis description.

STRUCTURED ANALYSIS

Products

The products are *data flow diagrams*, *minispecifications* expressed in structured English or pseudocode, *decision tables* and *decision trees*, and the *data dictionary*:

- *Data flow diagrams.* 'The purpose of a *data flow diagram* (DFD) is to show, for a business area or a system or a part of a system, where the data comes from, where the data goes to when it leaves the system, where the data is stored, what processes transform it, and the interactions between data stores and processes' (Gane, 1990).

  On the diagram, *sources* and *sinks* (sometimes termed *external entities*) show sources and destinations of data that are outside the system; that is they are objects about which we do not need to record information. *Data stores* represent places where data is stored in the system and *data flows* may be seen as pipelines through which groups of related data flow between data stores, processes and external entities. Finally, processes transform data in some way.

The technique termed *levelling* is used for dividing a data flow diagram into parts for easier readability, analysis or for implementation purposes, and the principle is that processes are refined into successive levels of detail. The top level is usually termed a *context diagram* and shows only the external entities, their input and output data flows, and the relevant system or organizational area drawn as a single process.

The next level, level 0, shows major sub-systems, external entities, data stores and data flow. The next level down, level 1, consists of a set of diagrams, where each diagram is a refinement (or 'explosion') of one of the level 0 sub-systems. Levelling of a process may continue until that process is a *functional primitive*, which usually means that its detailed process logic can be written in a page of structured English. Each diagram is numbered so that the relationships between diagrams and processes are easily traced. The principle of data conservation must also apply, which states that data at a lower level must be included in the data at a higher level.

A DFD aims to set a boundary to the system and is meant to be non-technical (Gane, 1990); that is it is meant to be understandable to business people who are familiar with the business area shown.

The relationships between the data and the processes are shown, but only on a simplified level. For example, exception or error processing should not be shown on a DFD, no timing (daily, weekly) is shown and important control structure such as sequence, selection and iteration of processes is also not shown. The DFD is meant as an overview of the required system.

- *Minispecifications.* These (also termed 'minispecs') describe the logic of functional primitives, that is the detail of the lowest level of processes from the DFD. Structured English is commonly used, incorporating sequence, selection and iteration. An example is given in Chapter 9.
- *Decision tables and trees.* These are used to diagrammatically illustrate complex conditions (selections) that are important and need checking by the user. Examples are given in Chapter 9.
- *Data dictionary.* The data dictionary (DD) is a central repository or 'encyclopaedia' which contains detailed components of an evolving specification. For structured analysis, it contains descriptions and definitions of data elements, structures, flows and stores.

  Each data flow from the DFDs is named and its components described, using data structure notations for hierarchical relationships, iteration and optionality. In addition, Gane and Sarson recommend that all data structures should be normalized into third normal form (see Chapter 8). This applies to data stores and files also. Relationships between files should be shown.

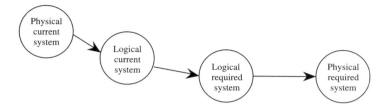

Figure 10.9 Recommended steps in structured systems analysis

Steps

Yourdon and de Marco suggest that one way to carry out analysis is to follow the steps shown in Fig. 10.9. These steps are based on the assumption that a new system will be based on an old system. The steps involve the initial production of what is termed a physical DFD. This is a DFD that models the physical information flow of the current system through the departments and individuals in the organization. Process names often refer to documents or machinery. The information may be on paper, over the telephone, and it may be used by people or a computer. Sometimes, the data flows may represent physical material flows, as well as information flows. This will be the case if we think we want to record information about such flows in the required system.

From this physical DFD, a logical DFD is produced in the second step, which abstracts the required data from its media, departments or individuals, and manual or computer systems. Step 3 takes place after various options for improving the current system have been considered and step 4 considers the necessary detail for producing a new physical design.

STRUCTURED DESIGN

Structured design applies to the physical design level and is mainly concerned with producing detailed program designs, which are implicitly for third-generation programming languages. Input is from structured analysis and also from the selection of target hardware and software. The main techniques are building structure charts, coupling, cohesion, transform analysis, transaction analysis and module packaging.

CASE STUDY

Data flow diagram

A logical data flow diagram for the required system is shown in Fig. 10.10. This is a levelled DFD, with three levels shown, which we shall now briefly explain:

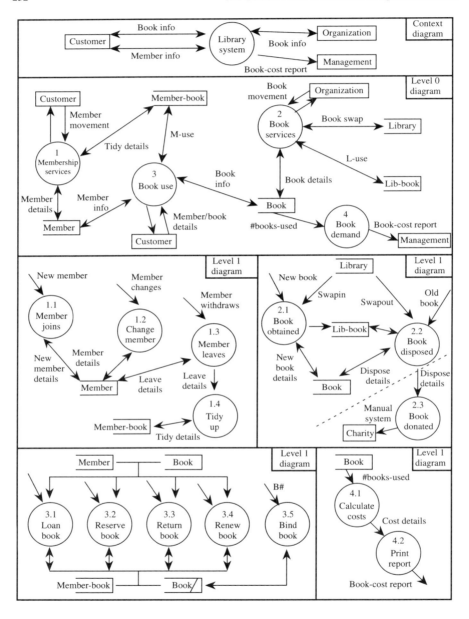

Figure 10.10 Structured analysis data flow diagram for library case study

- *Context diagram.* This shows the main system required (library system), external entities, which are outside the system, and information input and output to the external entities.
- *Level 0 diagram.* This refines the main system process from the context diagram to show the four main sub-systems of membership services, book services, book use and book demand. The five data stores are (a) member, which records information about all library members, (b) book, which records information about all books, including the number of uses and whether a book is being bound or has returned from binding, (c) library, which records information about all external libraries, (d) lib-book, which records information about which books have been swapped in or out with which external libraries, (e) member-book, which records information about all books a member loans, reserves, returns or renews.
- *Level 1 diagram.* There are four level 1 diagrams, one for each of the four processes on the level 0 diagram, and each diagram shows the refinement of its 'parent' process. The membership services sub-system is refined into members joining, changing membership details, leaving and tidying-up; book services consists of obtaining, disposing of and donating books; book use refines into loaning, reserving, returning and renewing books to or for members, and binding books; and book demand provides a report on book costs.

    Space does not permit us to show bind-in and bind-out processes, which are refinements of the bind 3.5 process, and similarly for the reserve-by-title and reserve-by-volume processes, as well as cancel reservation, which are refinements of the reserve book 3.2 process.
- *Human–computer boundary.* The boundary between the computer and the manual system is often indicated on the DFD. In this example, process 2.3 (book donated) represents the manual process whereby, after an unwanted book's record is modified, the book itself is donated to a local charity.

Minispecifications

The logic of functional primitive processes is produced in the minispecifications. In structured English:

```
Process 3.1
BEGIN
   receive 'book' info
   IF book does not exist
      THEN
         BEGIN
            #uses = 0
            .........
         END
END
```

The process would contain exception processing and error messages, such as when we try to lend a book reserved by another member.

Data dictionary

The data dictionary contains data elements, data structures, data flows and data stores, and we shall show an example of each. Allowed ranges of values, or actual values permitted, may also be shown. We use indentation to show hierarchical relationships between elements for data structure, flow and store. There will also be extensive cross-references between data as well as between data and process.

element	B#
alias	Bnumber
description	primary key of Book
data type	7 characters AN
range	B1000–B999 999
structure	Book-line
	B#
	Book-description
	#uses
flow	Book-cost report
	Heading-book-totals
	Book-line*
store	Book
	B#
	Btype#
	#uses

JSD

INTRODUCTION

The JSD method of system specification was created in the UK by Michael Jackson (Jackson, 1983) and has been added to subsequently. In the 1983 book, there are six steps: (a) entity action, (b) entity structure, (c) initial model, (d) function, (e) system timing and (f) implementation.

Our description is based on this publication, as well as Cameron (1986) and more recent JSD courses, which describe three main stages for creating JSD specifications: *modelling, network* and *implementation*. The correspondence between the steps and the stages is:

Stage	*Step*
Modelling	Entity action
	Entity structure
Network	Initial model
	Function
	System timing
Implementation	Implementation

The modelling stage consists of defining an initial model of the objects and processes in the real world that the analyst judges should form the core of the desired system, and then elaborating that model; major products are the model process structure, operations table and context error table. The network stage specifies the links of the initial model to the real world, including input and output processes and a human–computer interface, and the product is the SSD, or system specification diagram. Finally, the implementation stage describes how the network processes may be organized so that the processes and data map into a physical implementation environment. This is shown on the SID, or system implementation diagram.

MODELLING STAGE

This stage creates a *real-world model* which contains entities, actions, model process structures and operations, which are refinements of actions. The actions are all *update* actions on entities. The initial model is produced first, consisting of an entity list, an action list and model process structures, which are then refined into elaborated models.

Initial model

- *Entities, actions, attributes.* Firstly, a list is made of all relevant object types, termed entities. The next step is to list all the actions that can take place on these entities (in JSD terms, entities *perform* or *suffer* actions). Next, the analyst decides which actions are performed or suffered by which entities, and list all relevant attributes of both the entities and the actions.
- *Model process structures.* These describe diagrammatically the possible *sequence* of actions that can take place on an entity instance. A type of entity life history diagram is used, with the entity as the root node and the actions shown below it. Structures are built for all entities taking part in actions. It should be noted that (a) action sequence is explicitly modelled and is read from left to right, (b) the sequence of actions is on a *given*

instance of the entity and not on *any* *instance* of the entity. The significance of this will be discussed in the case study example.

In addition to the action sequence, selection and iteration are also shown, and Fig. 10.11 shows the three control components of model process structures. In the figure the box (*structure node*) labelled A is a *sequence node*. This is denoted by the occurrence of one or more unmarked nodes drawn below it, and the action sequence of these nodes is shown diagrammatically by left-to-right ordering. Node B is an *iteration node* and node C is a *selection node*. An iteration node is represented by having a single node below it which is marked with an asterisk. This node may execute zero or more times. A selection node is shown by the occurrence of one or more nodes below it, which are marked with a small circle. In the figure, one of E4 or E5 must execute, but not both.

There are no conditions associated with a process structure. The structure specifies what action sequences are possible and not the circumstances in which particular sequences occur. The E nodes in the diagram are termed *elementary nodes*.

The E6 node in Fig. 10.12 is a null node, allowing for no action to be taken at this point. The !B node shown is a *quit node*. This node will check for a condition and if it is true the node B is exited and the next node in the sequence is actioned. In this case the next node is C.

Quit nodes are used in process structures to exit a branch of the hierarchy typically for exceptions, such as when the 'normal' action sequence of an object may be suddenly terminated.

Elaborated model

- *Operations*. For each of an entity's actions, there is a set of *operations* that are performed on the entity attributes. The operations are refinements of

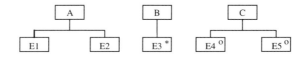

Figure 10.11   Sequence, iteration, selection and elementary nodes

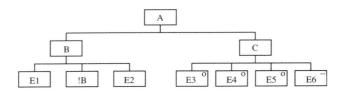

Figure 10.12   An example of a null and a quit node

the action and are drawn on model processes as boxes containing numbers. Each number represents one operation. The analyst will create a table specifying the operations against their *operation numbers*. The specification of operations is not precise, as it is written in natural language.

- *Resume points and text pointers*. Resume points are also shown in the process structure. A resume point is added to the start of each process structure and to the end of each elementary node. The eventual record that will represent the required attributes of each entity instance will also have a field called a *text pointer*. The text pointer records the resume point for the entity. This has an initial value of one. The text pointer is used to show the status of the entity instance with respect to its life history. The record of the attributes and text pointer of an entity instance is known in JSD as the *entity state-vector*.
- *Context errors*. At run-time, there may be errors in the input data to the actions of model processes or there may be certain rules, as we saw in Chapter 9, that must not be violated by the actions.

    In order that the action data may be checked, a *context error table* is created. This is used to define rules (pre-conditions) on the actions in which an entity can take part. A context error table defines errors for each entity, resume point, action and relevant condition.

NETWORK STAGE

This stage is where the model processes, modelling the real world, have input and output functions added to them and their logic is specified in detail. The chief product is the system specification diagram (SSD), which is a type of data flow diagram.

All processes have their logic specified in the same diagrammatic notation as process structures, as well as using structure text for detailed specification.

First step (initial model)

This consists of showing all the model processes (only the entity node is shown) on the SSD, together with the data that flows between them. The logic of the processes, including the operations, is also specified in more detail as *structure text*. Data flow is shown using the *datastream and state vector*:

- *Datastreams*. The possible sequence of action messages from the real world can be considered as forming a queue to the model process. This is a datastream. Datastreams may also be used to connect the model processes to other types of process, the *function processes*, discussed below.
- *State vector*. A process will often want to know the values of the attributes of an entity instance. These are known as the entity state vector.

Second step (function processes)

The SSD so far only contains processes modelling the core organizational activities with their data flows, so this step adds function processes, which are the input and output processes that are required to transmit data to and from the system, together with data flows. There are three types of function process, input, information and interactive:

- *Input processes.* These read any input data for the model processes (*action messages*) and check that it does not contain context errors, using the information stored in the context error table. As well as dealing with the context errors, input processes check that the attributes of action messages are of the correct type (for example date, integer). These errors are called *message errors* and are checked against information specified with the action attributes. User interface details are also specified.
- *Information processes.* An information process sends messages to the outside world. Sometimes a process will inspect the state vector of some other process. This state vector inspection is represented in a network diagram as a connection marked with a diamond, as in Fig. 10.13. The leaving members process produces a list of all members who have left since the last reporting time, by inspecting many instances of member state vector SV-1 and producing output on the DS-5 datastream.
- *Interactive processes.* A system may contain a function process that automatically generates action messages, input to model processes. For example, a stock control system might automatically generate orders for materials and parts.

Third step (system timing)

This step obtains information relating to time restrictions on the processes, such as the degree of synchronization between the real world and the system, affecting decisions to do with batch, on-line or real-time processing, the periodicity of reports (for example on request, daily, weekly) or time

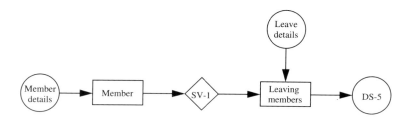

Figure 10.13   Information process, with state-vector inspection and output datastream

dependencies between processes themselves, such as terminal response time. Only narrative means are used to specify this.

CASE STUDY

Initial model

The entity list includes the following:

Entity	Definition	Attributes
Book	Books that are in the library	B#, Btype#, #uses
Member	Individuals who may use the library	M#, Mtype#
Library	External libraries who swap books	Lib#

The action list includes the following entries:

Action	Definition	Attributes
Lend	Someone borrows a book	B#, M#, date
Acquire	The library acquires a book	ISBN, title, Lib#, date
Join	An individual becomes a member	Name, address, date

Model process structures

An example is given in Fig. 10.14, where actions are represented by the leaves of the hierarchy. Join, renew and leave and so on are actions performed by (an instance of) the entity member. The actions on a process structure are all performed or suffered by the same real-world object and are thus mutually dependent. For example, in Fig. 10.15, renew depends on the book having been obtained and lent.

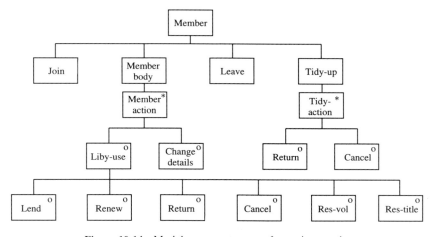

Figure 10.14 Model process structure for entity member

However, there may be another set of actions in which the book instance can take part. For example, the action res-vol (reserve book) is not constrained by whether the book is on loan or being rebound, which means that res-vol is independent of actions such as lend and should appear on a *separate* process structure. However, res-vol is dependent on swap-in, acquire, swap-out and sell and so these actions must also appear in this second process. The result is the process book avail (Fig. 10.15).

The two process structures are said to represent different *roles* of the entity book. This is the means by which JSD can model concurrent processes, as an action in the process book-avail (for example, res-vol) may be executing (on

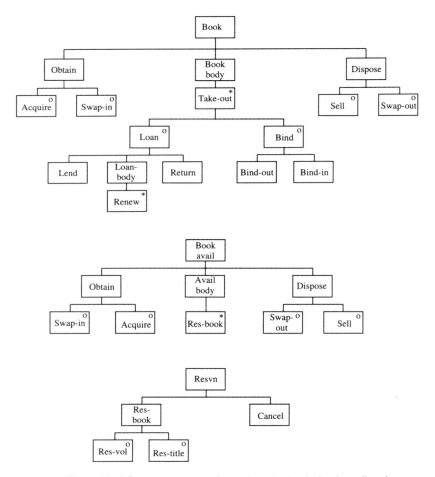

Figure 10.15 Process structures for entity roles book, book avail and resvn

the same instance of a real-world book) at the same time as an action (for example, return) in the process book.

Since res-vol appears on more than one process structure, it is said to be a *common action*. For the resvn (reservation) entity role, also shown in Fig. 10.15, all actions are common actions.

Elaborated model

- *Operations and resume points*. Figure 10.16 shows the book process structure with operations and resume points, and Fig. 10.17 shows part of a matching operations table. A book with a text pointer of 2, for example, will have just been acquired by the library (see Fig. 10.16), that is acquire would be the last action, of interest to the process structure, book, in which the book has taken part. The next actions that the book can take part in are lend and bind-out. If the book next suffers a lend (the book is lent to a member without having been reserved), its text pointer is changed to 4 and if the book suffers a bind-out the text pointer is changed to 7.
- *Context error table*. Figure 10.18 shows that for a book at resume point 6, receiving an action message relating to action lend, and under the condition that a member is trying to borrow the book which is reserved by another member, error 888, 'Book is reserved', will occur.

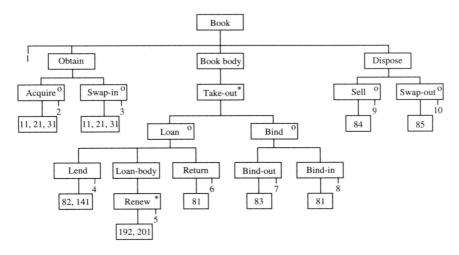

Figure 10.16 Process structure for entity role: book, with operations and resume points

Operations	
Numbers	Text
11	Store ISBN
31	Set library-no
201	Set member-no

Figure 10.17   A table of operations

Entity role	Resume point	Action	Condition	Error number/OK	Error note
Book	1	Acquire		OK	
Book	1	Swap-in		OK	
Book	1	Lend		1	Library does not own book
Book	2	Acquire		37	Book already obtained
Book	2	Lend	Lend-date ≥ obtain-date	OK	
Book	2	Lend	Lend-date < obtain-date	43	Lend earlier than obtain
Book	6	Lend	Book-reserved-by = nil	OK	
Book	6	Lend	Book-reserved-by ≠ nil	888	Book is reserved

Figure 10.18   Context error table

Network stage

- *Input processes.* These include the grouping of many actions for the convenience of the user. For example, a screen might allow the user to enter the details of many reservations being made by one member at once. Input processes specify all aspects of the human–computer interface.
- *Information processes.* An information process sends messages to the outside world. The network diagram in Fig. 10.19 shows a connection between an input process, the book model process and the inventory report information process to produce a book costings report. The process is reading the state vector SV-1 of book (obtaining values of #uses for each book instance) and writing to the datastream DS-5. Detailed logic of the inventory report process is specified in a process structure diagram and in structure text.

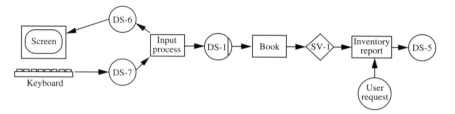

Figure 10.19 An input and information process connected to a model process

The representation of 'screen' and 'keyboard' are for illustrative purposes only. They are not standard JSD symbols. The vertical bar on datastream DS-1 indicates that the datastream is a *control datastream*, which is a datastream that may be both read and written by a given process and is 'locked' to write operations of other processes in between the read and the write.

- *Interactive processes*. An example of this process type is cancel function, which automatically cancels any reservations made by a member on leaving. Figure 10.20 shows that it sends messages to the resvn and member processes.

As only entity nodes are shown on the SSD, there would be many data-streams and state vectors representing data flows between the underlying but unspecified (on the diagram) entity actions. For this reason, the SSD is generally used only to show local parts of the network involving function and model processes.

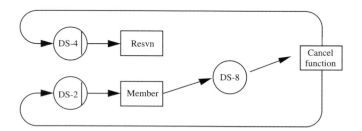

Figure 10.20 An interactive function — cancel function

SSADM

INTRODUCTION

The SSADM method (structured systems analysis and design method) is produced by the CCTA, a UK government agency (SSADM, 1990). The method originates from course material developed in the early 1980s by Learmonth and Burchett Management Systems plc (LBMS), with additional features.

The main variant of SSADM, intended for medium to large systems, will be described. The most recent version of SSADM is version 4, and the main phases and products (referred to as *modules* and *products*) may be seen in Fig. 10.21, compared to those of version 3. A guide to SSADM version 3 may be found in Downs, Clare and Coe (1988) and Longworth (1989). The

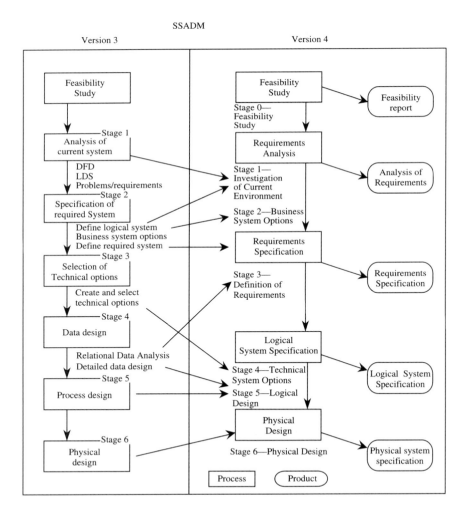

Figure 10.21   Comparison of version 3 stages and version 4 modules of SSADM

evolving requirements and specification are stored in what SSADM terms the *requirements catalogue*.

FEASIBILITY STUDY

Stage 0 — Feasibility Study

This assumes as a starting point that a project has been identified as the result of an exercise such as strategic planning, or something similar, and it sets out and evaluates different technical, organizational, financial and business options. The aim of the module is to establish whether the direction of the project and the requirements are feasible, and if it is therefore worth committing the necessary resources for project development. Feasibility study essentially consists of a shortened, higher-level version of the next two modules.

REQUIREMENTS ANALYSIS

This consists of two stages. In stage 1, requirements are defined by investigating the current environment and identifying problems or areas that need improvement. Stage 2 then develops a range of options that meet the defined requirements and selects one option as the basis for the desired system.

Stage 1 — Investigation of Current Environment

The stage begins by creating an overview of the processing and data in the current environment (within the framework set by the feasibility report), and then documents problems, necessary improvements or new data or functions required. The intended users of the new system are also identified.

Subsequent steps analyse the processing and the data in more detail, finally producing a logical view of current processing. Data and processing are analysed in parallel, and the logical data flow diagram (DFD) and the process descriptions should refer only to entities. Products used are:

- *Data flow diagram.* This is the main product used for describing processes, and only update processes are shown. Enquiry or report processes will be shown later on the *enquiry access path* (see below). Physical DFDs are produced initially, which are then transformed to logical DFDs after the data has been analysed. The lowest level processes are described in narrative form on the *elementary process description*.

  A maximum of three DFD levels are recommended, with further process detail being shown in function definitions and entity–event modelling (see below). Exception or error handling should not be shown on the DFD.

- *Logical data structure.* This consists of a set of entities and relationships obtained by analysis of the data in the current system, using the entity modelling technique. The logical data structure (LDS) may also contain the most significant attributes.

Stage 2 — Business System Options

In this stage, several feasible high-level options for a proposed system are presented, which typically encompass the range of requirements from the mandatory to the optional. It is suggested that up to six 'skeleton' business system options are produced by the developers, with two or three being presented to the users. The selected option becomes the foundation for subsequent requirements specification.

A business system option (BSO) consists of a description of a proposed information system. It will typically be couched in terms of the existing environment, and each option should address the functional requirements, in terms of boundary, inputs and outputs, and principal transformations.

In addition, each option should also contain a description of how it will meet non-functional requirements, covering, for example: (a) priority and impact — the problems and requirements at present are noted together with the priority of the proposed system and its potential impact on the organization, for example, to reduce library processing delay for a loan from four minutes to one minute might have the highest priority; (b) costs and time — costs and time of development and operation are estimated, as well as costs of hardware and software procurement and training; (c) technical points such as volume and volatility of expected data storage, as well as estimates of task frequencies. This will also cover options for handling processes on- or off-line, as well as for batch as opposed to real-time processing.

The system boundary is an important feature as this will be drawn to distinguish computer and manual processes.

REQUIREMENTS SPECIFICATION

Using the option selected by business system options, a detailed specification of requirements now begins. The emphasis is on determining the desired system data, functions and events. Prototyping techniques are also suggested for the development of the human–computer interface.

Stage 3 — Definition of Requirements

The first two steps modify the previously defined DFD and LDS (which were of the current system only) to match the requirements in the selected BSO. All the attributes are specified on the LDS. In addition, non-functional requirements such as security, access and archiving requirements are defined.

The next step is to define *functions*. This involves identifying both update and enquiry functions and determining the events that are related to update functions. The input and output data for each function is then defined, using the *input/output structure* (see below). As most systems have on-line processing, system dialogues are identified in outline. As a check on the LDS, some functions have their input and output data analysed by *relational data analysis*, and the resulting relations are compared with the LDS entities.

The next step suggests prototyping the requirements with users to identify errors and to obtain any additional requirements. Procedures for managing prototyping sessions are provided and dialogues and report formats are emphasized.

Finally, using *entity–event modelling*, more detailed processing requirements are obtained. This is done by constructing an *entity life history* (ELH) for each entity from the LDS, and an *effect correspondence diagram* (ECD) is constructed for each event, showing the entities affected by that event. An *enquiry access path* is created for each enquiry showing the entities on a subset of the LDS that are to be accessed.

Methods used are:

- *Function definition*. A function 'is a set of system processing that the user wishes to carry out at the same time to support his business activity'. This technique builds functions from the DFD and enquiry processing requirements from the requirements catalogue. It then combines these and subdivides processes into separate functions. Functions are specified on forms.

- *Input/output structure*. This is based on JSP structure diagrams (Jackson, 1975), with data structure being shown in terms of sequence, selection and iteration. It may also be used to show on-line input and output dialogue data.

- *Relational data analysis*. This technique consists of the application of normalization to the input/output structures defined above to produce a set of third normal form relations. The main aims are to validate the LDS, to ensure that the data is logically easy to maintain and extend and to group the data together into optimum record types. The relations are a basis for defining databases or files in logical system specification.

  It is a bottom-up technique, as opposed to top-down entity modelling, and it may also be used in stage 1, where the input will be current system files, input/output documents and screens.

- *Specification prototyping*. This produces a live, 'dummy' system with which the user can experiment to refine the requirement for a user interface or to add other requirements.

- *Entity–event modelling*. This is used to define the DFD processes in more detail by including the effects of time on the system. This is done by considering entity events, where an event is 'something that triggers a process to update the system data'. *Entity life history analysis* constructs an ELH for each entity on the LDS, showing the updating events in the sequence in which they occur. Detailed operations are also added to each ELH as well as state indicators, which are equivalent to JSD resume points. The *effect correspondence diagram* takes the opposite view to the ELH and analyses the entities that are affected by each event. JSD diagrammatic notation is used.

• *Enquiry access path.* This shows, on a subset of the LDS, the entities and relationships involved in accessing the information required by an enquiry. The entry point is also shown. It uses JSD notation to indicate data structure.

LOGICAL SYSTEM SPECIFICATION

Stage 4 — Technical System Options

This assesses the different options for implementing a part of the specification and describes options, costs, benefits and constraints. Factors include internal and external constraints. External constraints consist of, for example, time, cost, business performance and any hardware or software restrictions set in the feasibility study. Internal constraints are, for example: (a) responsiveness — the responsiveness of the system is decided, considering synchronization issues such as real-time or periodic (monthly, yearly) and enquiry types, which might be *ad hoc* or scheduled; (b) sizing — numbers of entity instances give file sizes and process sizes, process frequencies are determined as well as number of lines in reports, and numbers of updates and reports; (c) security; (d) interfacing to other systems.

The aim is to select the best set of technical products that meet the requirements. A *technical environment description* for the chosen options is input to logical design.

Stage 5 — Logical Design

Dialogue design produces a design of the interface for on-line functions. Included are the dialogue specifications, command and menu structures. *Logical database process design* uses information from entity–event modelling to construct *update process models* and the enquiry access paths are used to construct *enquiry process models*. Only non-procedural specifications are produced in this stage. Processes have integrity checking and error processing added, and are consolidated into processing structures corresponding to an event.

PHYSICAL DESIGN

Stage 6 consists of physical design, and a classification framework is provided for types of physical processing and database management systems, enabling certain set procedures to be followed when translating logical designs to physical designs specific to one of these types of implementation environments. In addition, the logical design may need to be refined to add any detail necessary for the physical environment. The techniques specified for this module are only very general, as it is intended that individual users will apply their own project-specific physical design techniques. This is to reflect the increasing use of packages and 4GLs.

CASE STUDY

Requirements Analysis

- *Business System Options*. The result of stage 2 might be that, to reduce costs, the library will not develop a totally computerized system to begin with, but will have a manual system for dealing with swap-ins and swap-outs (a card-index filing system could be decided upon later for recording book numbers in and out), as well as binding. Member services will be a batch system (additions and changes made at the end of the day), with the central book processes on-line. A decision has been made to keep only three months of data on-line at a time, using a simple archiving system. The book donated process will be excluded from the system.

Requirements Specification

- *Data flow diagram*. Figure 10.22 shows levels 1 and 2 of the logical DFD of the case study produced in stage 3. The context diagram was set in stage 1. The diagrammatic notation uses rectangles for processes and ovals for external entities. In addition, as the DFD shows only processes that update data stores, and not enquiries or reports (which simply move data around the system), process 4 (book demand) is omitted. A decision was taken in business system options to exclude the donation of a book from the system. The diagram is, in other respects, identical to that of structured system analysis.
- *Logical data structure*. The logical data structure is an entity model of the type discussed in Chapter 8. Figure 10.23 shows this, which is similar to the final data map built for Information Engineering, but which does not show entity subsets directly. However, disjoint subsets, such as bind-in and bind-out, as well as res-title and res-volume, are shown as exclusive 1 : 1 relationships, indicated by an arc over the relationships. Only entities and their relationships are shown on the diagram, with attributes and relationship detail documented separately. Relationship names are normally present on the figure but have been omitted to avoid diagram clutter.
- *Cross-checking*. At this point, the DFD and the LDS are cross-checked, to make sure that important data has not been omitted and that there is

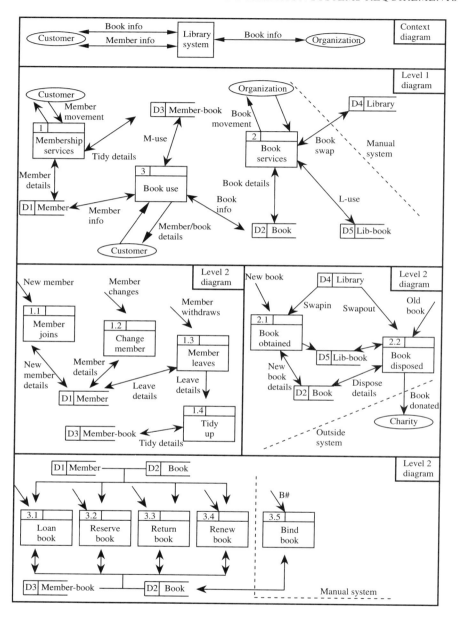

Figure 10.22 SSADM data flow diagram for case study process descriptions

correspondence between entities and data stores. The *logical data store/ entity cross-reference* documents the entities involved in each data store. For example, the swap-in and swap-out entities on the LDS correspond to the lib-book data store, and details about loans, renewals, returns and reservations are stored in member-book.

The *process/entity matrix* documents, for each process, the type of access (update or retrieval) made to each type of entity. For example, the bind-in and bind-out types of book are updated by the bind book process 3.5.

• *Relational data analysis.* The set of normalized relations corresponding to the LDS in Fig. 10.23 is shown below. Subset relations are shown by a corresponding *type* relation, containing a distinguishing attribute. The date attribute of the book relation is set when a book is sent to or received from binding.

Relation	Attributes
Book-type	B-type#, b-description (bind-in/out)
Book	B#, Btype#, bind-date, #uses
Lib	Lib#
Member	M#
Swap-in	Lib#, B#, date
Swap-out	Lib#, B#, date
Loan	M#, B#, date
Renewal	M#, B#, date
Return	M#, B#, date
Reservation	M#, B#, Rtype#, date
Res-type	Rtype#, r-description (title/volume)

• *Enquiry access path.* If we assume that a simple enquiry for the number of uses of a book will satisfy the requirement for information to help with determining book costs, then an enquiry access path for obtaining this information is shown in Fig. 10.24. This shows that the key attribute of the book entity, B#, is used as the entry point and that a set of books is to be retrieved, indicated by the iteration symbol in the book box. Each book entity contains the attribute #uses.

• *Entity life history.* Figure 10.25 shows the ELH for entities book, resvn and member, the first two of which have the same content as those in the JSD case study. However, one important difference is that in SSADM the ELH models *events*, while in JSD the process structure models *processes*. We may see the implications of this in the figure, for in the ELH for the member entity the tidy-up process is omitted, as it is not an event.

Another aspect of ELH specification that differs from JSD is that which is concerned with common actions and parallel processes. To model the fact that some member processes are in common with book processes (for

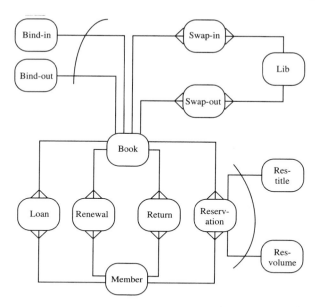

Figure 10.23 SSADM LDS for case study

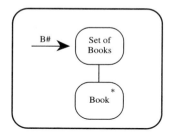

Figure 10.24 Enquiry access path for book use enquiry

example lend), JSD includes book actions on the member process structure (Fig. 10.14) as selections. SSADM simplifies this, as is shown in Fig. 10.25, by showing a book event on the ELH for member.

The requirement that a book may be reserved at any time, outside of the chronological event sequence of loan, renew and return, does not require a separate process as in JSD (see Fig. 10.15 for the process book-avail), but is modelled on the ELH for member as a parallel structure with book, shown by the double interconnecting line on Fig. 10.25.

• *Effect correspondence diagram.* Figure 10.26 shows the effect correspondence diagram for the lend event, where all entities affected by this event

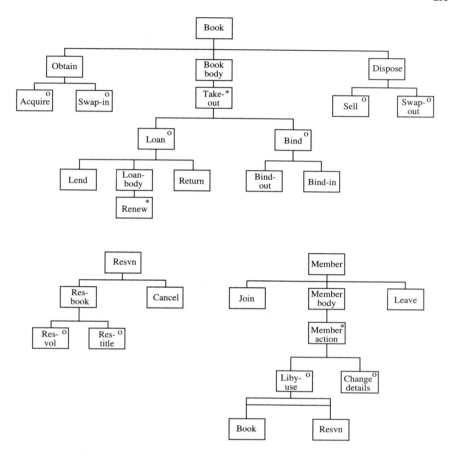

Figure 10.25 Entity life history for book, resvn and member

are detailed. The double-headed arrow between entities indicates that when one entity is updated, so is the other. Also shown is event data, which consists of the attributes input to the update process — in this example, member and book numbers M# and B#.

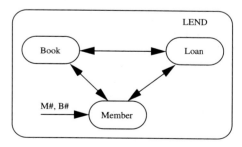

Figure 10.26 Effect correspondence diagram for lend event

Comparison of methods

It is not our intention here to initiate an extensive comparison of methods, and fuller treatments may be found in other authors (Olle *et al.*, 1988; Avison and Fitzgerald, 1988; Maddison *et al.*, 1983; Connor, 1985). In addition, our case study is chiefly suited to comparing methods for their model-building activities, and does not contain details that would enable us to compare them for other important method features such as validation and verification methods, strategic planning coverage, aids for user participation and so on.

However, the case study does allow comparison in terms of the *breadth* and *orientation* features. By breadth we mean the extent of the systems development phases covered by the method. This also considers necessary activities within the phases. For example, an analysis phase should have activities for modelling object, rule, process and HCI, as defined in Chapter 8. From the breadth we may determine the phase that has the most emphasis within a method, for example, requirements centred or logically centred. In addition, the orientation of a method is concerned with that aspect of the requirements, such as data or process, for which most support is provided.

We will use the phases and steps that we established in Chapter 6 for the comparison.

RESULTS OF COMPARISON

Information Engineering (IE)

This covers three phases, strategic planning, analysis and logical design (slightly), but only the object and data steps of each, and it is thus planning/conceptually centred. There is a lot of assistance for applying normalization and planning, and as IE basically produces what is often termed a corporate data model, it is a planning/data oriented method. IE also has a project management phase.

Structured systems analysis (SSA)

This only covers the analysis and logical design phases, considering process as well as data, but is logically centred, as data and not entities are considered and the emphasis is on process detail. There are some techniques (such as decision tables and trees) provided for process, but little for data, so this is process oriented.

JSD

The model process structure part of JSD may be considered to be on the analysis level, while elaboration is in logical design. However, objects and data are considered even less than in SSA, and JSD is logically/conceptually centred. The emphasis is firmly on the detail of process design and JSD is thus very process oriented.

SSADM

This covers the feasibility study part of requirements determination, the analysis phase (object, event and process) and logical design (data, process, HCI), so this is conceptually centred. There is some assistance for HCI modelling. There are many techniques (LDS, ELH, DFD) provided, as well as detailed assistance, so the method is data, event and process oriented.

Table 10.1 summarizes this comparison. The advantage of using a practical example for comparison, as opposed to a more theoretical exercise, is that the similarities between the methods are clearly visible. It is hoped that the conclusion may be drawn that, where they cover the same steps in the same phases, they model the same thing in roughly the same way.

Future method trends

Early methods were mostly concerned with the design level, and modern methods aim for a wide breadth, aiming to model as much as possible on the conceptual level, or higher levels, for the reasons explained at the beginning of this chapter. It is thus likely that methods will evolve into being *requirements centred*, emphasizing the requirements determination phase (and possibly strategic planning). This may involve prototyping trial systems to users from a specification.

Early methods were primarily process oriented and it is to be expected that methods will increase their set of orientations as they evolve. This means that they will provide modelling assistance for particular aspects of the user requirement and will be able to construct and present it in a variety of

Table 10.1 Breadth of coverage of IE, SSA, JSD and SSADM

Method	Strategic planning	Requirements determination		Analysis					Logical design		
		Feasibility	Analysis	Object	Rule	Process	Event	HCI	Data	Process	HCI
IE	X			X					X		
SSADM		X		X		X	X	(X)	X	X	X
SSA						X			X	X	
JSD						X			X	X	

different ways. For example, future methods are likely to become *object oriented*, modelling a greater variety of objects on the conceptual level, and are likely to include the modelling of organizational rules on this level.

Growing use of methods, and their growing breadth, is also likely to bring with it increasing demand for *contingent* methods, which include techniques for determining the best way to tailor the method to a particular set of application circumstances. Guidelines already exist for using different types of prototyping, for example.

Many aspects of current methods evolved by trial and error in commercial use. However, research methods are likely to play a part in the future for evaluating features of methods, such as new specification languages or modelling techniques. Some work has already been done in this area on HCI techniques.

In the future, organizations are likely to use methods to an increasing degree, and when the benefits of having an organizational portfolio of specifications are realized this will lead to method standardization. This is already happening with SSADM in the United Kingdom and to an extent with the Euromethod project in the European Community. However, a major force for standardization will probably be due to CASE tools which support systems development. This topic forms the basis of the next chapter.

Problems addressed and solutions provided

The two major problems addressed by modern methods are the problems of quality and productivity. As we have discussed, a quality problem occurs when systems do not do what users require and a productivity problem is when systems are delivered late or over budget.

QUALITY

The main problem that methods address is the third quality problem, concerned with making errors in the analysis of information needs. Many delivered systems are not used and another large proportion require significant amendment, as they do not do what users thought they would do, because the designers made mistakes of this type. Two reasons for this are:

1. *Poor development methods*. Designers may make errors when developing systems, due to lack of knowledge of suitable methods and techniques, as well as reliance on intuition. They may not know how to look for such typical errors, quoted in Davis (1990) as incorrect facts, omissions, ambiguities, inconsistencies and misplaced facts which occur in user requirements. In addition, there are inadequate methods for validating requirements with users, still mainly based on showing diagrammatic or narrative specifications.

2. *Designers cannot communicate*. Systems development is a team effort, requiring many designers to communicate together. The absence of

commonly understood phase products or activities means that mistakes and misunderstandings may be made.

Both of these are compounded by the sheer complexity of large systems.

The solutions provided are, firstly, for modern methods to emphasize the early phases more, including requirements determination and analysis, as this is where most errors are made, as we saw in the discussion in Chapter 6. Modern methods thus tend to be conceptually centred.

Secondly, knowledge has grown concerning more effective activities in the development process, including modelling techniques, approaches to validation and many others. By including this body of knowledge in a method, designer intuition may be reduced, so that fewer errors will be made in developing systems. Modern methods provide guidance and precise modelling techniques in an endeavour to capture more accurately, in a set of requirements, what users want.

Thirdly, there is a move towards methods that provide standard phase products, expressed in well-defined languages, or that use well-documented aids such as tables, matrices, diagrams and so on. If all the members of a development team are familiar with these expressive means then their ability to communicate together is enhanced.

PRODUCTIVITY

Problems

1. *Changing requirements.* Requirements may change after the project has started, causing work to be redone. A major reason for this is that *users do not know their requirements.* Users may not fully appreciate the nature of the system that the analysts are proposing to build for them, as they may be unsure about what they want. Alternatively, there may be disagreement over requirements between users. In addition, analysts can only express the nature of the eventual system in a computer-oriented way, mostly in the form of detailed diagrams, which do not allow users to fully comprehend their implications.

2. *Poor project control.* Many projects have no means of their status being measured at any given point, in relation to the amount of work completed or remaining to be done. Techniques for estimating required resources at the beginning of a project are inadequate, often giving rise to over-optimistic predictions for delivery dates and costs.

Solutions

1. *Allow users to refine their requirements.* Methods should not expect requirements to be fixed at the start, and it is suggested that, as users will never understand computer-oriented specifications, even after education, it is better to validate the requirements built by the analysts by using

prototypes to demonstrate a part of the eventual system to the users and to allow them to learn their requirements in more detail. This feature is beginning to appear in methods, and SSADM version 4 contains suggested procedures for prototyping part of the HCI to users.

2. *Develop project control techniques.* The phase products defined in a method may be used as milestones to chart project progress, accompanied by sizing and estimating techniques, perhaps incorporated in a project management phase closely coupled to the method, to estimate required resources on a changing basis more correctly. There is some evidence of progress being made towards standardization in the United Kingdom in the emergence of the PRINCE project management method (CCTA, 1990).

It is also hoped that the improvements in quality resulting from the better definition of modelling and validation techniques mentioned above will reduce the amount of iteration required in the process, and hence reduce the productivity problem also.

ADVANTAGES OF METHODS

The main feature of modern methods is that they are based on the application of the principle of *abstraction* to the systems development process, concentrating on the early phases. They also emphasize their products, defining them more precisely than before, as well as providing assistance for phase activities. They may be compared to previous methods, for example, structured methods, which emphasized activities. Method advantages include:

1. The development process is less complex as it is decomposed into distinct, simple phases. Each phase is concerned with one level of abstraction only, and its well-defined activities map the user requirement progressively from that level to the next. Guidance is provided for the activities, reducing designer intuition. Project control is also assisted.

2. The existence of high levels of abstraction in the early phases means that the users can be involved more fully in the development process, using validation and prototyping, as the semantics of user requirements are not embedded in implementation detail. This can reduce error in determining requirements.

DISADVANTAGES OF METHODS

1. Methods may only address that part of the problem concerned with agreed documentation and modelling procedures. There are other problems (see Chapter 6), which may be caused by poor methods for defining the problem or assessing wider organizational issues, or the lack of talented development people.

2. Early phases of the process have more emphasis and so add to resources required at development time (although this should decrease maintenance effort).

Summary

This chapter began by defining the notion of a method and introducing some common terminology. A brief chronological history of methods was then outlined, progressing from the pre-method era, through structured, data-oriented and future methods.

Four methods, Information Engineering, Structured systems analysis and design, JSD and SSADM, were then compared for the way in which their model-building activities dealt with a simple case study. A comparison framework was introduced in terms of method breadth and orientation, and some future trends were considered. The problems addressed by methods and the solutions they provide were then discussed, followed by advantages and disadvantages.

Discussion questions

1. What were the main trends that marked the change from the time referred to here as the 'pre-method era' to structured methods?
2. What problems were addressed by the move from structured to data-oriented methods?
3. List the three most important problems, in your opinion, that methods address. Why do you select these?
4. Compare the DFDs from SSA and SSADM. What are the differences, if any?
5. Compare the data map of IE with the output from relational data analysis of SSADM. Is there any overlap?
6. Compare the entity life history in JSD to that in SSADM for the case study example. Which of the two is easier to understand and why?

References

Arthur Young Information Technology Group (1987) *The Arthur Young Practical Guide to Information Engineering*, Wiley, Chichester.

Avison, D. E and G. Fitzgerald (1988) *Information Systems Development: Methodologies, Techniques and Tools*, Blackwell Scientific, Oxford.

Baker, F. T. (1972) 'Chief programmer team management of production programming', *IBM Systems Journal*, vol. 11, no. 1, pp. 56–73.

Bjorner, D. (1987) 'On the use of formal models in software development' in *Proceedings of the 9th International Conference on Software Engineering*, Monterey, CA, pp. 17–29.

Boehm, B. W. (1975) 'The high cost of software' in *Practical Strategies for Developing Large Software Systems*, E. Horowitz (ed.), Addison-Wesley, London, pp. 3–14.

Bohm, C. and G. Jacopini (1966) 'Flow diagrams, Turing machines and languages with only two formation rules', *Communications of the ACM*, vol. 9, no. 5, pp. 366–371.

British Standards Institute (1987) BS 5750, Part 1: *Specification for Design/ Development, Production, Installation and Servicing*, Part 2: *Guidance*. Available from British Standards Institute, Sales Department, Linford Wood, Milton Keynes, MK14 6LE.

Cameron, J. R. (1986) 'An overview of JSD', *IEEE Transactions on Software Engineering*, vol. SE-12, no. 2, pp. 222–240.

CCTA (1990) *PRINCE: Structured Project Management* (5 vols), NCC Blackwell, Oxford.

Chen, P. P. (1976) 'The entity–relationship model: towards a unified view of data', *ACM Transactions on Database Systems*, vol. 1, no. 1, pp. 9–36.

Codd, E. F. (1970) 'A relational model of data for large shared data banks', *Communications of the ACM*, vol. 13, no. 6, pp. 377–387.

Connor, D. (1985) *Information System Specification and Design Road Map*, Prentice-Hall, Englewood Cliffs, NJ.

Daniels, A. and D. Yeates (1969) *Basic Training in Systems Analysis*, Pitman, London.

Davis, A. M. (1990) *Software Requirements: Analysis and Specification*, Prentice-Hall, Englewood Cliffs, NJ.

de Marco, T. (1979) *Structured Analysis and System Specification*, Prentice-Hall, Englewood Cliffs, NJ.

Dijkstra, E. W. (1968) 'Go To statement considered harmful', *Communications of the ACM*, vol. 11, no. 3, pp. 147–148.

Downs, E., P. Clare and I. Coe (1988) *Structured Systems Analysis and Design Method*, Prentice-Hall, London.

DTI (1988) *Software Quality Standards: The Costs and Benefits. A Review for the Department of Trade and Industry*, April, Price Waterhouse. Available from Library and Information Centre, Department of Trade and Industry, 1–19 Victoria Street, London SW1H 0ET.

Eason, K. (1988) *Information Technology and Organisational Change*, Taylor and Francis, London.

Fagan, M. E. (1976) 'Design and code inspections to reduce errors in program development', *IBM Systems Journal*, vol. 15, no. 3, pp. 182–211.

Finkelstein, C. (1989) *An Introduction to Information Engineering*, Addison-Wesley, Wokingham.

Friedman, A. L. and D. S. Cornford (1989) *Computer Systems Development: History, Organization and Implementation*, Wiley, Chichester.

Gane, C. (1990) *Computer-aided Software Engineering*, Prentice-Hall, Englewood Cliffs, NJ.

Gane, C. and T. Sarson (1979) *Structured Systems Analysis: Tools and Techniques*, Prentice-Hall, Englewood Cliffs, NJ.

Horowitz, E. (ed.) (1975) *Practical Strategies for Developing Large Software Systems*, Addison-Wesley, London.

Humphrey, W. S. (1988) 'Characterizing the software process: a maturity framework', *IEEE Software*, vol. 8, pp. 73–79.

Jackson, M. A. (1975) *Principles of Program Design*, Academic Press, London.

Jackson, M. A. (1983) *Systems Development*, Prentice-Hall, Englewood Cliffs, NJ.

Kenny, A. D. (1989) *Managing Software: the Businessman's Guide to Software Development*, Blackwell Scientific, Oxford.

King, D. (1984) *Current Practices in Software Development: A Guide to Successful Systems*, Prentice-Hall, Englewood Cliffs, NJ.

Lee, B. (1979) *Introducing Systems Analysis*, NCC Publications, Manchester.

Longworth, G. (1989) *Getting the System You Want: A User's Guide to SSADM*, NCC Publications, Manchester.

Maddison, R. N., G. Baker, L. Bhabuta, G. Fitzgerald, K. Hindle, J. Song, N. Stokes and J. Wood (1983) *Information System Methodologies*, Wiley, Chichester.

MOD (1991a) 00-55: *The Procurement of Safety Critical Software in Defence Equipment*, Part 1: *Requirements*, Part 2: *Guidance*, April. Available from Directorate of Standardisation, Ministry of Defence, Room 0140, Kentigern House, 65 Brown Street, Glasgow, G2 8EX.

MOD (1991b) 00-56: *Hazard Analysis and Safety Classification of the Computer and Programmable Electronic System Elements of Defence Equipment*. April. Available from Directorate of Standardisation, Ministry of Defence, Room 0140, Kentigern House, 65 Brown Street, Glasgow, G2 8EX.

Myers, G. T. (1975) *Reliable Software through Composite Design*, Petrocelli-Charter, New York.

Nassi, I. and B. Schneiderman (1973) 'Flowchart techniques for structured programming', *ACM SIGPLAN Notices*, vol. 8, no. 8, pp. 12–26.

NATO (1981) AQAP-13: *Software Quality Control System Requirements*, August. Available from Ministry of Defence, DGDQA, Room 134, Building 22, Royal Arsenal West, Woolwich, London, SE18 6ST.

NATO (1984) AQAP-14: *Guide for the Evaluation of a Contractor's Software Quality Control System for Compliance with AQAP-13*, May. Available from Ministry of Defence, DGDQA, Room 134, Building 22, Royal Arsenal West, Woolwich, London, SE18 6ST.

Naur, P. and B. Randell (eds) (1968) *Software Engineering*. Report of a conference sponsored by the NATO Science Committee, Garmisch, Germany.

Oakley, B. (1988) 'The issues of today' in *Software Engineering (Proceedings of Software Tools 88 Conference, London*), Blenheim Online Ltd., London, pp. 1–4.

Olle, T. W., H. G. Sol and C. J. Tully (eds) (1983) *Information Systems Design Methodologies: A Feature Analysis*, North-Holland, Amsterdam.

Olle, T. W., H. G. Sol and A. A. Verrijn-Stuart (eds) (1982) *Information Systems Design Methodologies: A Comparative Review*, North-Holland, Amsterdam.

Olle, T. W., H. G. Sol and A. A. Verrijn-Stuart (eds) (1986) *Information Systems Design Methodologies: Improving the Practice*, North-Holland, Amsterdam.

Olle, T. W., A. A. Verrijn-Stuart and L. Bhabuta (eds) (1988) *Computerized Assistance during the Information Systems Life Cycle*, North-Holland, Amsterdam.

Olle, T. W., J. Hagelstein, I. G. Macdonald, C. Rolland, H. G. Sol, F. J. M. Van Assche and A. A. Verrijn-Stuart (1988) *Information Systems Methodologies: A Framework for Understanding*, Addison-Wesley, Wokingham.

Orr, K. T. (1977) *Structured Systems Development*, Yourdon Press, New York.

Ross, D. T. and K. E. Schoman (1977) 'Structured analysis for requirements definition', *IEEE Transactions on Software Engineering*, vol. SE-3, no. 1, pp. 6–15.

Royce, W. W. (1970) 'Managing the development of large systems' in *Proceedings of IEEE WESCON*, pp. 1–9. (Reprinted in *Proceedings of the 9th International Conference on Software Engineering*, March–April 1987, Monterey, CA, pp. 328–338).

Royce, W. W. (1975) 'Software requirements analysis: sizing and costing' in *Practical Strategies for Developing Large Software Systems*, E. Horowitz (ed.), Addison-Wesley, London, pp. 57–72.

SSADM (1990) *SSADM Version 4 Reference manual* (4 vols), July, NCC Blackwell, Oxford.

Stevens, W., G. Myers and L. Constantine (1974) 'Structured design', *IBM Systems Journal*, vol. 13, no. 2, May, pp. 115–139.

Warnier, J-D. (1974) *Logical Construction of Programs*, Van Nostrand Reinhold, New York.

Yourdon, E. (1978) *Structured Walkthroughs*, 2nd edn, Yourdon Press, Englewood Cliffs, NJ.

Yourdon, E. (1982) *Writings of the Revolution: Selected Readings on Software Engineering*, Yourdon Press, Englewood Cliffs, NJ.

Yourdon, E. and L. L. Constantine (1979) *Structured Design: Fundamentals of a Discipline of Computer Program and Systems Design*, Prentice-Hall, Englewood Cliffs, NJ.

11
Tools

Introduction

This chapter discusses the use of automated tools in systems development. Generally defined, a tool provides software support for precisely defined aspects of systems development methods. There are many terms for these tools, such as: modelling tools, software tools and, more recently, CASE tools, where CASE is an acronym for computer-aided systems or software engineering.

METHOD AREAS AND TOOLS

There are many tools that are commercially available, representing a fast-growing segment of the software market. They differ widely in their functionality, and there are tools for the technical activities of systems development, as well as for related areas such as strategic planning, configuration management and project management.

For the technical activities, there are tools for diagramming (for example, entity modelling, data flow diagrams), code generation in the implementation phase, user interface modelling, prototyping and reverse engineering in the maintenance phase. So-called upper or front-end CASE tools support activities in the early phases (these are also termed analyst workbenches or analyst toolkits), while lower or back-end CASE tools support the later phases.

Early tools such as STRADIS/DRAW, introduced in 1981, were only diagramming tools, providing drawing and printing facilities. In comparison, modern tools allow specification objects (for example, objects such as entity or process) to be represented graphically and to be stored in a database, in such a way that the objects may be independently accessed by related tools, which may wish to use them for purposes such as cross-referencing or syntax checking. This feature excludes traditional development software, such as compilers, from being regarded as CASE tools.

TYPES OF TOOLS

There are four types of tool, from the viewpoint of coverage of systems development. Firstly, single function tools aim to automate only one aspect of the development process, such as data flow diagramming. The second type consists of a family of tools, which typically cover more than one phase and which may be obtained singly or in combination. These do not aim to support any particular method, leaving the designer to choose what tools to use within a project. The individual tools generate products with a common format, which are stored in a database used by all the tools. Output from one tool may thus be input to another.

Thirdly, I-CASE (integrated-CASE) environments consist of a tightly integrated set of tools using a common database, which may or may not support a specific method and which increasingly cover several or all of the phases. One of the first tools of this type was EXCELERATOR, introduced in 1984, which was mainly based on the Yourdon approach to structured systems analysis. More recent I-CASE tools cover all phases and offer code generation in the implementation phase.

The above three types may all offer the designer the possibility of a limited amount of tool tailoring. For example, it is often possible to define your own type of diagram.

The fourth type of tool, generally termed an integrated project support environment, or IPSE, is a more advanced type, emerging from research and development, and is intended to integrate project and configuration management tools with development tools. Other aims are flexibility, in that facilities are provided to allow designers to configure their own method and method support, and integration, allowing designers to import tools into the IPSE and use them in an integrated manner. However, not all products claiming to be IPSEs offer such comprehensive facilities.

We shall examine AUTO-MATE PLUS in detail, which is an I-CASE environment from LBMS in common use in the United Kingdom, and we shall cover the analysis and design phases only. We shall also briefly look at Systems Engineer, also from LBMS, which is representative of a newer generation of I-CASE environments.

I-CASE architecture

The architecture of an I-CASE environment consists of a specification database, which is used to hold a specification, accompanied by various software tools which, for analysis and design, help designers to build, check and display the specification in different ways. Such an architecture may be seen in Fig. 11.1. Other terms for the specification database are data dictionary, data inventory, repository, encyclopaedia and design database.

The specification database is often in two sections, as shown, one containing components which are in the process of being built, the other containing

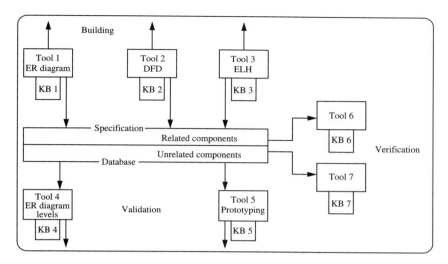

Figure 11.1 I-CASE architecture for analysis and design phases

those which have been completed, and which may be related to other completed components.

TOOLS AND KNOWLEDGE BASES

The figure shows different tools surrounding the central specification, each tool having access to its own knowledge base (KB), where information is kept specific to the tool function. For example, the knowledge base for tool 1 will know, at least, the allowable syntax of the ER diagram. If the tool is more advanced, its knowledge base may contain information that will allow the tool to advise or assist the designer in building the ER diagram.

The diagram does not show, for the sake of clarity, that knowledge bases may be partially shared by tools. For example, tool 6, which checks ER diagram syntax, will also have access to the same syntax knowledge as tool 1.

Depending on tool implementation, the specification database may only be available for one user at any one time, although more recent tools allow many users to be updating the specification concurrently.

TOOL FUNCTIONALITY

The architecture also provides a way of describing tools based on the type of activity they carry out:

1. *Building tools.* These allow the specification to be built and are mainly used by the designer. They are likely to be highly interactive tools.
2. *Verification tools.* These are for activities that can be almost completely automated. They check properties of the specification. For example, they

may check (a) the syntax of various components of the specification (for example, an entity relationship without a name), (b) the consistency between different but related parts of the specification (for example data in DFDs and relations).

3. *Validation tools.* These tools are used by both user and designer and are intended to check that the specification meets the user requirements. Ideally, they may present parts of the specification in a variety of different forms, most suited to those wishing to examine the specification. For example, users may wish to have an overview of detail in the specification, to see a prototype or to see data flow diagrams in Gane and Sarson as opposed to de Marco formats. Similarly, designers may wish to focus on just a part of a specification and see this in relation to other parts, for example to examine all entities in the 'library services' area together with related processes.

It may not be easy to separate validation activities from those of building, particularly when in contact with a user, as validation will often bring to light more parts of the requirement. Some environments may allow the designer to switch easily between validation and building tools, or for requirements or designs to be changed immediately, when in a prototyping session with the user.

Problems and solutions

We shall use the problems discussed in the methods chapter to understand the solutions tools attempt to bring and then briefly mention some disadvantages. As for methods, it is generally held that no one technique or tool can, by itself, bring improvements of an order of magnitude to quality and productivity, such as increasing productivity ten times. Instead, we should expect steady improvements to follow from the joint application of a variety of tools.

QUALITY — SYSTEMS DO NOT DO WHAT USERS REQUIRE

As for methods, the main problem addressed by tools is the third quality problem, concerning the errors made in analysing information needs. The main two problem causes are:

1. *Poor development methods*
2. *Designers cannot communicate*

The solutions that tools provide to this problem are:

1. *Tools provide more effective techniques.* There is some evidence to show that tools are more successful at developing systems that meet the technical aspects of user requirements. This may be due to the increased analytical effort required, as tools provide more effective modelling techniques, for example, entity models and entity life histories, as well as the availability of validation techniques such as prototyping tools. They

also make certain types of error detection more feasible, as they provide the automatic means to detect, for example, incomplete or inconsistent requirements more easily. They also provide assistance to aid (or supplant) analyst intuition, such as automated help, or techniques such as on-line entity modelling syntax checking.

2. *Tools define phase products.* A tool provides well-defined phase products, as well as techniques and approaches that all analysts will use. Training in the use of the tool will be required, thus enabling analysts to communicate more effectively.

3. *Tools may offer code generation.* The automated generation of code from a logical design avoids the possibility of programmers making programming errors. However, it should be borne in mind that, as stated in Chapter 6, coding errors only account for approximately 7 per cent of total errors. In addition, tools do not currently (despite the use of the term 'complete') generate all the code that is required.

A productivity implication is that experienced staff are freed from mundane tasks and can concentrate their expertise on more difficult areas.

PRODUCTIVITY — MANY SYSTEMS ARE DELIVERED LATE AND OVER
BUDGET

The problems are:

1. *Changing requirements*
2. *Poor project control*

The solutions that tools provide are:

1. *Requirements are refined with prototyping aids.* Tools can communicate more effectively with users, compared to paper-based diagrams or narratives, as they make it possible to use prototyping or specification execution techniques, or presentation methods tailored to the individual user, to demonstrate a system to users, allowing them to refine their requirement.

2. *Machine-readable products.* Precisely defined phase products that are machine readable may assist the evaluation, estimation and configuration control tasks within project management. In addition, defined procedures and products are more visible and thus project slippage can be more easily detected. These are theoretical benefits, as there is no evidence yet that they have been realized.

Those who advocate the benefits of tools identify three further factors, addressing the manual nature of the development process, which are held to contribute to the productivity problem:

3. *Avalanche of paper.* Developers may spend a lot of time keeping track of paper documents containing parts of the specification. A tool reduces the

paper document problem considerably as the specification is machine readable.

4. *Detailed and error-prone work.* Much of the work involved in systems development, particularly in the design and implementation phases, is very detailed and often tedious, requiring large amounts of time. A tool can automate some aspects of the implementation phase, by partial code generation (for example, user-computer interaction or code for individual transactions), and can also, by presenting standard options or applying standard techniques, reduce the amount of detail that has to be supplied or reduce the time spent by the designer (for example, optimizing a set of relations or comparing two entity models).

5. *Tracing between related parts of the specification.* The problem of changing requirements causes difficulties in keeping all related parts of a specification up to date. Impact analysis and global change facilities, built on a tracing facility, help to reduce this problem. Improvements in these 'mechanical' features may also reduce errors and so help the quality problem.

It has recently been claimed (Macdonald, 1990) that, for the I-CASE environment information engineering facility (IEF), the improvement over manual methods for productivity is 3 : 1 and the improvement for quality is 3.7 : 1. However, findings conflict, as research in manual COBOL environments (Jeffery and Lawrence, 1985) showed that technical changes, such as on-line development and testing, did not provide productivity improvements, although such improvements were one of the major reasons for adopting this type of change. Instead, productivity improvements were correlated with the management and technical ability of programmers' supervisors. A survey (HECTOR, 1990) shows that users perceived tools to bring some improvement in quality, but that productivity was not significantly affected.

DISADVANTAGES

Some disadvantages are, according to the NCC STARTS Guide (NCC, 1987):

1. *Costs.* There are extra costs, for example, staff training costs, consultant costs to help with method or tools and purchase/maintenance costs of tool hardware and software. In practice, it has not been found that development costs are significantly reduced using tools. Although benefits should come in the long term with reduced maintenance, as the system is easier to change, a faster payback on tool investment may be required.

2. *Quality.* Although small quality improvements have been reported, most tools cover only the routine tasks of systems development, and it is difficult for tools to replace human experience in areas such as detection of ambiguity or the identification of unreliable or critical areas of the specification.

3. *Productivity*. It has not been found that productivity is significantly improved for the analysis and design phases, possibly because tools require a much more rigorous, and hence time-consuming, approach to be adopted for these phases, compared to the time taken when manual methods are used.
4. *Risks*. The wrong choice of tools may be made. Many vendors exist, and the chosen one might go out of business. Alternatively, an organization might end up with out-of-date tools, as methods are currently in a transition period.
5. *Incompatibility with other aids*. Tools do not integrate well with other development aids, such as 4GLs, methods or tools already being used. More recent versions of tools are beginning to address this problem.

AUTO-MATE PLUS

INTRODUCTION

AUTO-MATE PLUS (LBMS, 1988) dates from 1985 and covers the analysis, logical design and physical design phases of the development process. It consists of:

1. Automated tools, which use diagrams and forms for building and viewing the specification.
2. Imported techniques, such as entity modelling, normalization and data flow diagrams.
3. An underlying software tool for tasks such as storing, checking and cross-referencing components of the specification.
4. A related method (non-automated), providing steps and advice.

The specification is on two levels. When a specification component is considered by the user to be complete, it can be automatically verified and then stored in the *design database*, where it can be linked to other parts of the specification also in the database. Prior to this, automated tools will store components outside the design database. The design database is single user only.

The description we shall give refers to Version 3 of the tool. The related method, LSDM/AM (LBMS Structured Development Method/AUTO-MATE PLUS), has been under development by Learmonth and Burchett Management Systems (LBMS) since the early 1980s and has formed the basis from which SSADM has evolved. The main (top-level) menu may be seen in Fig. 11.2.

OVERVIEW

Table 11.1 shows an overview of the phases and steps of the related method. Those aspects of the method with tool support are shown by the name of the

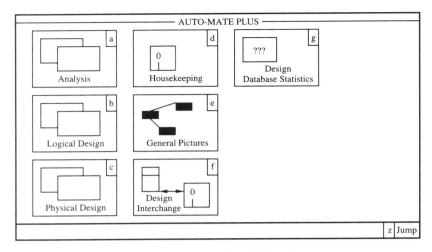

Figure 11.2   AUTO-MATE PLUS main menu

relevant phase product in the appropriate column. The D/F column indicates whether the tool uses diagrams (D) or forms (F) for its user interface.

The related method is only slightly integrated with the tools, because, for example, the tools do not have to be used in the sequence of the method steps. Only in logical design, when applying relational data analysis to produce the TNF, is guidance given to the user.

1. *Analysis phase.* In this phase, there are diagrammatic tools for producing the LDS (logical data structure), DFD and on-line dialog, an abstract model of an on-line user interface. Other tools are forms based.
2. *Logical Design phase.* Relational data analysis uses normalization to produce a set of relations in third normal form (TNF). This is merged with the LDS from the analysis phase to produce the composite logical data design (CLDD). A diagrammatic technique is used to specify DFD processes in more detail with the entity life history (ELH), error handling being described in narrative on forms.
3. *Physical Design phase.* In the last phase, for the data part, generation tools produce database definition statements from the CLDD. The process part defines process outlines, consisting of physical transactions, modules and low-level process statements.

We shall describe the two main types of user interface for tool interaction, pictures and forms, and then discuss the detail of the three phases.

Table 11.1 AUTO-MATE PLUS: related method, software support and diagram/form for user interface

Method task list and steps	Software support	D/F
Analysis		
A1 Initiate the project	—	—
A2 Investigate data structure (current system)	LDS	D
A3 Investigate current system processes	DFD	D
A4 Identify problems with current system	Problem/requirements list	F
A5 User review	—	—
A6 Identify user requirements	Problem/requirements list	F
A7 Identify business systems options	—	—
A8 Further define the chosen option		
Process	DFD	D
System function	System function	F
Event	Event	F
Data	LDS	D
Data item	Data item	F
On-line dialog	IDS/TDS	D
Screen formats, I/O formats	—	—
Prototype on-line dialog	Dialog test	—
A9 User review	—	—
Logical design		
B1 Relational data analysis	TNF	F
B2 Create final logical data design	CLDD	D
B3 Define system logic	ELH	D
B4 Finalize logical design	Error handling	F
B5 User review of logical design	—	—
Physical design		
C1 'First cut' physical data design	PDS	F
C2 Create process outlines (major transactions)	Process outlines	F
C3 Create performance predictions	—	—
C4 Create process outlines (minor transactions)	Process outlines	F
C5 Complete physical design tasks	—	—

PICTURE EDITOR

Introduction

The *picture editor* is used to create and maintain all the different types of diagrams, for example, logical data structures, data flow diagrams and entity life histories. The basic screen used for drawing pictures is similar to all diagram types, and a simulation of this screen, for drawing data flow diagrams, may be seen in Fig. 11.3.

The screen is divided into four areas:

1. *Heading.* Prompts and messages from the system appear here.
2. *Window.* This is where the picture being edited is shown.

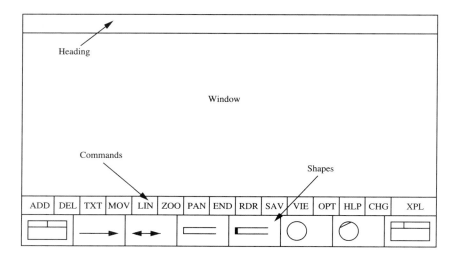

Figure 11.3   AUTO-MATE PLUS picture editor — DFD screen

3. *Commands.* Separate boxes contain text commands for use in editing a picture. These are selected before selecting an icon or link symbol (shape).
4. *Shapes.* Boxes contain the different icons and links that make up the type of picture being edited.

The three basic types of picture object are icons, links and text. *Icons* represent objects and the tool will prompt for associated text, such as a name. *Links* join icons, representing flow direction, icon relationship or transition, and also have text associated with them, which is again prompted for. Finally, *text* may be entered anywhere in the picture, using the TXT command, as well as being associated with links or icons.

Commands

Although commands and shapes will vary between screens for different picture types, many commands are common. For example, the HLP command gives help information, SAV saves the current picture in a picture file and END usually terminates the editing session for the current picture, returning to the menu.

It is often the case that a picture is larger than the window area of the screen, and the PAN command allows the user to move the window over other parts of the picture. This is done by selecting part of the picture to be the new window centre. To look at part of a picture in more or less detail, the ZOO command allows the user to zoom in on a picture, for greater detail, or to zoom out, for an overview.

The MOV command allows single or multiple objects to be moved from one place in the picture to another. If there is a link between a moved object, then the editor will attempt to redraw the link as a straight line. If this is not possible, then the user will be prompted for intermediate points on the link.

Selecting picture objects

With a mouse, selection is accomplished firstly by moving the mouse so that the arrow (cursor) on the screen is in the required position and then by pressing the right-hand mouse button. Depending on what is required, the arrow may point at a command, icon or link, or at an object or point on a picture.

Picture operations

Pictures are saved in picture files which are separate from the design database. All picture types have a range of menu options for copying, deleting, printing and having their contents listed in reports. In addition, the load/delete menu option will check the picture syntax (termed *validation*) and, if correct, load the picture into the design database. Otherwise, error messages will be provided.

FORMS MANAGER

Some techniques, for example, relational data analysis, are not supported by the picture editor. Instead, the *forms manager* maintains different types of screen-based forms, containing headings and fields which, assisted by system prompts, are used to obtain the necessary information from the user and store it in the design database. Reporting facilities are also available to display such information in the design database.

Unlike picture-based validation, where all of a picture is checked at once, forms-based validation checks each item as it is entered. Items may be committed to the design database at any time when using a form. Customized forms may be designed.

ANALYSIS PHASE

The analysis menu may be seen in Fig. 11.4. We shall assume, when discussing the tools within the different phases, that the relevant (analysis, logical design, physical design) menu has been previously selected.

Logical data structure

- *Introduction.* The picture editor is used for creating the LDS diagram, a type of entity model. The picture allows entities, relationships and 1 : *n* relationships to be shown. Subsets are not shown and attributes (termed

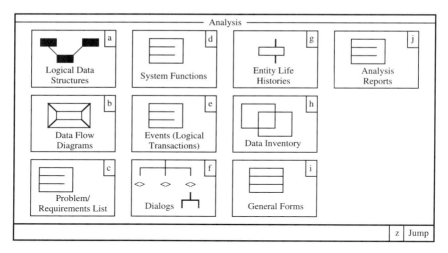

Figure 11.4   AUTO-MATE PLUS analysis menu

data items) are stored in a separate, associated, component. The user selects the logical data structures option from the analysis menu in Fig. 11.4.

● *LDS picture editor*. The picture editor for LDS is entered by selecting the edit LDS picture option from the LDS menu. After the user replies to the prompt for an LDS name, the screen comes up, similar to the DFD screen in Fig. 11.3, but with the shapes shown in Fig. 11.5. A 'crow's foot' option is possible instead of the arrowed line.

   — *Create an entity*. An LDS picture might be started by drawing an entity. This is done by firstly selecting the ADD command. Next, the required icon is selected (the entity icon is the rectangle) by moving to the correct box on the shapes lines shown in Fig. 11.5 and pressing the mouse button. The mouse is then moved until the cursor (which is now a grey box shape) is in the required position in the window area; the mouse button is then pressed. The icon will be positioned at this point and the system prompts in the heading line for the entity name, which the user enters via the keyboard and which is then shown inside the icon.

   — *Add a relationship*. After ADD is selected, the correct type of link icon is selected. Figure 11.5 shows a range that allows optionality and a rule termed exclusivity to be shown on the link. The heading line prompts FROM?, to which the user replies by selecting the icon from where the link should begin. The prompt now says TO?; the user selects a destination icon and the system draws the link. To name a relationship, the TXT command is used. The position on the link is selected and the user is prompted for the name.

Figure 11.5   Shapes line for edit LDS in AUTO-MATE PLUS

— *Save.* To save an LDS in a picture file, the SAV command is selected and the LDS is given a name.
• *Adding further detail.* To add more detail to an LDS, the menu option entities, reports, relationships is selected, which shows a third-level menu, and from this, options are selected for entities, relationships, or *data items* (attributes).

Screen forms are available to enter information, which will not be shown on the LDS but will be available in a report. For entities, detail stored may be description, number of instances and number of insertions in a given period. For relationships, information includes description, role name and cardinality (if 1 : 1). For data items, detail includes related entity(s), name, role, description, optionality, key data or not, length, data type and format, and synonyms.

Data items may also be entered into the *central data item inventory* within the design database by selecting the data inventory option from the analysis menu and then the maintain data item definitions option from the next menu.

Data flow diagram

For data flow diagrams, the user selects the data flow diagrams option from the analysis menu, giving the DFD menu.

• *DFD picture editor.* The picture editor for DFDs is entered by selecting edit DFD picture from the DFD menu. Level 1 is the highest level DFD (there is no context diagram). The basic editing operations are drawing processes, data flows, data stores and external entities. These icons may be seen on the shapes line in Fig. 11.3, and the procedure is similar to that for drawing the LDS.
  — *Exploding DFDs.* A feature possessed by DFDs is that they are levelled, in that a given process may be refined into many (AUTO-MATE PLUS allows six) levels of detail. To do this, the XPL command is first selected and then the required process. Subsequent action depends on whether the process has previously been exploded. If so, the screen changes to the DFD on the next level of detail of the process. If not, then a blank process box is shown, with all related icons from the higher level shown outside the box, which may then be used to create this lower level of DFD. The END command is used to return to a parent DFD.

— *Other features*. DFDs may have more detailed (narrative) process descriptions and cross-references between LDS entities and data stores entered by the user.

Problem/requirements list

This allows the forms-based entry of current system problems and corresponding solutions, constraints or requirements for the new system. The user selects the problem/requirements list option from the analysis menu. As many of these forms may be completed as required, one for each problem. No cross-referencing is possible and all information is in narrative form.

System functions

A system function is a type of computer processing that is characterized by factors such as batch or on-line, timeframe (on request, daily, monthly) and origination (depot, head office). The user selects the system functions option from the analysis menu. System functions are defined, using a form, and DFDs or DFD processes are assigned to one or more of the functions. A system function should be associated with only one level 1 DFD process.

Event

An event, which is a logical transaction, triggers update or enquiry processing, and events are inferred from DFDs and LDSs by intuition. The user selects the events (logical transactions) option from the analysis menu and a form appears, allowing an event to be defined with its name and the related system function, DFD, or DFD process.

On-line dialog

- *Introduction*. This feature allows the design of interactions between the user and an on-line system. Such interactions are made up of menus, transactions and control flows, often related in a hierarchical manner. For example, interaction with the AUTO-MATE PLUS system constitutes an on-line dialog, beginning with a top-level menu, allowing the user to select a path through sub-menus and eventually transactions and to branch to various points as desired.

  Two levels of detail are provided. The first level consists of a framework of menus and transactions which constitute a particular dialog; this is termed an *invocation dialog structure* (IDS). The second level expands the detail of any one menu or transaction into a set of structured *exchanges*, which may consist of a set of screens; this set is termed the *transaction dialog structure* (TDS).

The tool provided consists of a detailed version of a state transition diagram. A state may be a menu, transaction or event. Control flows are labelled with the conditions that cause control to flow in that direction, branch symbols are provided which allow GO TOs to be specified in the structure and other symbols are provided allowing for global setting of control values, condition descriptions and setting control values on or off.

- *Invocation dialog structure (IDS)*. Selecting the dialogs option from the analysis menu and then selecting the edit IDS picture option invokes the IDS picture editor. Figure 11.6 shows an IDS picture for the library case study. This shows a dialog consisting of a main menu with three options. Selecting M takes the user to the member menu, selecting B to the book-use menu and selecting Q terminates the dialog by returning from the main menu to the invoking environment.

Details of the member menu are not shown, but the book-use menu has five options. Selecting L passes control to the loan transaction, while Ren, Ret and Res each pass control to their respective transactions. Selecting B causes control to be transferred 'back' to the menu or transaction which is one level above, in this case, the main menu.

Other features are the two global settings and resultant actions of the Clear and PF 10 keys. If the Clear key is selected, at any point in the dialog, the dialog will be terminated. If the PF 10 key is selected, control will return 'back', with the context determining where control flows. This can be used after each of the lowest-level transactions, for example.

The icons shown in Fig. 11.6 are all available on the shapes line in the IDS picture editor screen. Another editing feature that is used is the JMP command. This is used when the window is too small for all the components in a dialog, and the procedure is to select JMP from the command line and then select one of the lower level menus. A new screen is generated with the selected menu at the top, ready for more icons to be attached.

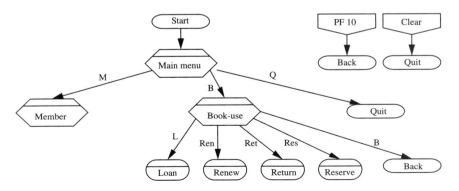

Figure 11.6 IDS picture for library case study

The notation is on an abstract level and does not specify, for example, how selections are to be made. The mouse could be used to select options on the screen or they could be entered from the keyboard.

- *Transaction dialog structure (TDS)*. The detail of each menu or transaction is defined here, including the number of screens, their sequence and data processed for each transaction or menu. Although most menus normally consist of only one screen, some transactions may need more than one (for example, a normal and an error message screen). Processing associated with a screen is termed an *exchange*. The TDS picture editor is invoked by selecting the XPL command from within the IDS picture editor and then selecting a menu or transaction icon. A new screen is generated with the previously selected icon at the top. The screen can only be specified abstractly and can only refer to the names of data processed. There is an auto-generate option, which will generate a single exchange TDS for each relevant IDS icon.
- *IDS dialog tester*. After an IDS has been loaded into the design database, it can be tested using the IDS dialog tester. The user selects the dialogs option from the analysis menu and then selects the dialog tester option. The name of the start-up menu or transaction is displayed, as well as the options associated with that menu or transaction as a selection list. Any option may be selected and the tester follows that path to the next state.

LOGICAL DESIGN PHASE

This phase adds further detail to the requirements specification from the analysis phase to produce a system logical design. It consists of tools for relational data analysis, entity life history and error handling narratives. In addition, results from relational data analysis and the LDS in the analysis phase may be combined into the CLDD. We assume that the user has selected the logical design option from the main menu shown in Fig. 11.2.

Relational data analysis

The technique used by this tool is normalization, and the aim is to input unnormalized data and transform it into a set of normalized relations or logical database records, without redundancy, termed TNF. The data may originate from screen layouts, reports and input documents of the required system or file layouts from a current system. The user selects the relational data analysis option from the logical design menu, which gives a set of options ranging from the input of raw relations to the merging of an LDS with a set of normalized relations.

The approach breaks down into several distinct stages. Firstly, raw relations are manually entered, giving relation and data item names. The next stage is to convert the raw relations into unnormalized relations by designating the data items which are key fields. At this point, unnormalized relations form the starting point of a *relation tree*, which is used to trace the normalization results on any given relation backwards or forwards.

The unnormalized relations may now be normalized up to third normal form. This is done by asking a set of questions about each relation, for example, 'are there any repeating groups [Y/N]' or 'are there any transitive dependencies [Y/N]'. Novice and expert modes are provided, and the user selects any data items that are identified by the questions. A backout facility is available during the process to remove relations from the tree or to return to a particular normalization state. When a set of third normal form relations has been produced, these may be optimized, to produce the smallest possible set.

Finally, an LDS may be generated from an optimized relation set and a *composite logical data design* (CLDD) may be created by merging an optimized set (or its generated LDS picture) with the LDS picture obtained from entity modelling.

Entity life history

- *Introduction*. To specify processing in more detail, this tool models the events (logical transactions) that affect the life of each entity. Only events that *update* entities and their data items (and entity relationships) are concerned (that is insert, modify, or delete). Events are those identified in the analysis phase and entities are from the CLDD, determined above.

  The tool supports the SSADM form of ELH diagram, which it terms hierarchical ELH, and its own form, which it terms network ELH and which is described below. The ELH cannot be loaded into the design database and so cannot be verified or cross-referenced to other parts of the specification. To enter the ELH picture editor, the user selects the entity life histories option from the logical design menu and then the edit ELH picture option.

- *Network ELH*. The notation for network ELHs uses a rectangle for an entity, a square for events that insert an entity, a circle for events that modify an entity and an inverted triangle for events that delete an entity. It is possible to show iterated events as well as selected events. However, the sequence in which events may occur for a given entity is shown more explicitly than for other types of ELH, as arrowed edges are used to show event sequence rather than a left to right ordering. The effect is rather like that of a flowchart, and Fig. 11.7 shows an ELH for the resvn entity from the library case study.

  The selection between res-vol and res-title is shown by the branch in the line after resvn, which can only be travelled once in the direction of the arrow (unless an event returns the life to an earlier state). The arrow

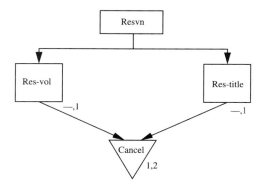

Figure 11.7   Entity life history (network) for resvn entity

shows that either res-vol or res-title have to occur before cancel (there are no modify events for this entity).

— *State indicators.* State indicators (similar to JSD resume points) may also be set for each event, using the TXT command, as shown in Fig. 11.7. They consist of two sets of values, the first set being all the valid state indicator values at the start of this event and the second the 'set to' value, to be set when the event terminates.

Error handling narratives

These are similar to JSD context errors, in that error conditions may arise when an event occurs and the entity is not in one of the valid previous states. For example, an error arises if a cancel resvn event occurs and the state indicator is set to null, meaning a resvn has not been created.

Error handling narratives are defined with forms, obtained by selecting the entity life histories option and then the maintain error handling narratives option. These allow the ELH valid and invalid states and consequent actions to be described, in narrative format.

PHYSICAL DESIGN

For the data part, a '1st cut' physical data design is produced by generating file structures or database definition statements (the physical data structure, or PDS) for a range of DBMS products, using the CLDD from logical design. A diagram of the underlying data structure can also be produced and edited.

The process part of physical design consists of forms-based assistance for specifying processes in more detail. The first step is to identify physical transactions, using the logical transactions defined in the analysis phase, in the on-line dialog, and in the entity life history in the logical design phase.

An IPSE path may also allow foreign tools to be imported. However, this requires the tools to be IPSE compatible. Compatibility means that they should share similar object concepts (such as entity, process, data item) and that they can interface downwards with the IPSE specification database and upwards with the IPSE user interface.

A degree of downward compatibility is in sight at present with the emergence of a European (PCTE, Portable Common Tools Environment) and US (CAIS) standard. However, the level of granularity supported is not sufficiently fine grained to allow, for example, two DFD diagramming tools which support the PCTE interface to read each other's specification database records successfully. The standards also allow for a distributed development environment, and it is likely that users and developers at remote locations will be supported by a distributed specification database, as well as tools to assist them in the process of working together.

Two other areas in which standards are emerging are the Information Resource Dictionary Standard (IRDS) for the specification database and the CASE Data Interchange Format (CDIF).

TOOL CHARACTERISTICS

Tools, which cover most, if not all activities, from strategic planning and project management to reverse engineering, are likely to become more common. Reverse engineering tools are likely to become particularly important, as they are aimed at the maintenance phase, which has been neglected until recently, and they will reduce the large applications backlog that holds up systems development in many organizations and is estimated to consume up to 90 per cent of system budgets in the 1990s (Oakley, 1988).

As both forward and reverse engineering tools are used, organizations will begin to build, on a specification level, precise descriptions of all their systems. This will enable specifications to be reusable for new applications, something already happening where corporate data models exist.

The application of formal methods such as Z (Spivey, 1988) and VDM (Jones, 1986) to safety-critical applications (such as nuclear reactor monitoring) is also expected to increase in the future. This will involve the expansion of the rather basic toolsets available at present, and relates to the reliability notion of quality.

An area that appears overdue for change is that of program design, where techniques such as structure charts have been used since the mid-1970s, and tools to support object-oriented design and programming techniques are likely to increase. As discussed under class 3 and class 4 tools, tools that have access to specialized knowledge bases or tools that can perform reasoning over the specification may also appear in the near future.

Summary

This chapter briefly discussed the main features of CASE tools, describing four main types and introducing an I-CASE architecture for analysis and design. The problems addressed by tools were then discussed and the solutions they provided. The main topic was a detailed description of the AUTO-MATE PLUS CASE tool. We then identified four main classes of tools, based on the difficulty of the task they were attempting to automate, and then presented some current tools. We finished with some guesses for the future of automated tools. Further reading, in an area that is quickly out of date, may be found in Rock-Evans (1987, 1989) and Williams (1990).

It is possible, as discussed earlier, that CASE tools result in an improvement in the quality problem, but it is not agreed that the productivity problem is significantly addressed. It will be necessary to wait for some years, until system maintenance is well under way, before it will be clear whether systems are, as claimed, more easily maintained.

Discussion questions

1. What is the essential difference between development aids such as compilers and assemblers, and CASE tools?
2. What are the four types of CASE tool? Do they all have different advantages and disadvantages?
3. Bearing in mind the problems addressed and solutions provided by CASE tools, summarize the three main advantages and disadvantages.
4. How does the AUTO-MATE PLUS method LSDM/AM fit into the systems development approaches discussed in Chapter 6? Would you classify it as structured, data oriented or requirements centred, as defined in Chapter 10?
5. Examine the differences between the AUTO-MATE PLUS and Systems Engineer products. Which do you think are the most important and why do you think they were made?

References

Gane, C. (1990) *Computer-aided Software Engineering*, Prentice-Hall, Englewood Cliffs, NJ.

Heap, G. (1988) *Evaluating Software Tools for Systems Analysis and Design*, NCC Publications, Manchester.

HECTOR (1990) *HECTOR Market Assessment UK Country Report*, HECTOR Esprit project number 2082. Available from KPMG Peat Marwick McLintock, 8 Salisbury Square, London EC4Y 8BB.

Jeffery, D. R. and M. J. Lawrence (1985) 'Managing programming productivity', *The Journal of Systems and Software*, vol. 6, no. 1.

Jones, C. B. (1986) *Systematic Software Development Using VDM*, Prentice-Hall, London.

LBMS (1988) *AUTO-MATE PLUS User Guide* (3 vols). Available from Learmonth and Burchett Management Systems, Evelyn House, Oxford Street, London.

LBMS (1990) *LBMS Systems Engineer User's Guide*, Parts 1 and 2. Available from Learmonth and Burchett Management Systems, Evelyn House, Oxford Street, London.

Macdonald, I. G. (1990) 'The impact of an integrated CASE system on the Information Engineering methodology' in *Proceedings of the 2nd International Conference on Information Systems Developers Workbench*, S. Wrycza (ed.), Gdansk, Poland, pp. 88–106.

NCC (1987) *The STARTS guide*, 2nd edn, NCC Publications, Manchester.

Oakley, B. (1988) 'The issues of today' in *Software Engineering (Proceedings of Software Tools 88 Conference, London)* Blenheim Online Ltd, London, pp. 1–4.

Rock-Evans, R. (ed.) (1987) *Analyst Workbenches*, Pergamon Infotech, Maidenhead.

Rock-Evans, R. (1989) *CASE Analyst Workbenches: A Detailed Product Evaluation*, Ovum Ltd, London.

Spivey, J. M. (1988) *The Z Notation — A Reference Manual*, Prentice-Hall, London.

Williams, R. (ed.) (1990) *Using CASE Tools in Systems Development: their Scope and Value*, Gower Technical, Aldershot.

Government reports

- A study (NEDO, 1983) in the United Kingdom of 15 organizations, reported in Winfield (1991), showed that only about half of the organizations found job reductions after computer systems were introduced. In some cases, many computer-oriented jobs were created.
- A report concerning the investment decisions of organizations in computer technology (Kearney, 1985) revealed that up to 20 per cent of the investment might be wasted, for example, software paid for but never delivered or software delivered but never used.
- A Department of Trade and Industry report (DTI, 1985) found that time-scale overruns occurred in 66 per cent of projects, while 55 per cent were over budget.
- For 'front-office' applications, a 1986 survey (DTI, 1986) showed that, of 20 pilot office automation projects begun in 1982, only half worked as the users expected, were fully accepted or brought positive benefits to the organization, and 20 per cent were rejected.
- A recent study (DTI, 1988) estimates that poor quality software costs £1 billion annually to UK companies, in making corrections before and after delivery, overruns and high maintenance costs. These are direct costs and do not include indirect costs such as lost business, lower effectiveness, damage to reputation or missed opportunities.

Management consultant surveys

- A survey of 252 companies (KPMG, 1990) showed that 'runaway systems' (systems with poor quality and productivity) concerned over 30 per cent of all major projects. The major effects of these systems were loss of time, reduction in staff morale, loss of money, reduced customer satisfaction and a negative market image.

Academic research

- Harrington (1991) describes research into organizations in the North East of the United Kingdom, where 60 per cent of the firms had major problems within two years of a computer-induced change in their structure.
- In a survey of 10 small companies setting out to implement computer systems (Wroe, 1985), it was found that only four managed to proceed to the implementation phase.
- Finally, Eason (1988) refers to studies in the United Kingdom in the 1980s which indicate that up to 40 per cent of systems may never be delivered, or may fail and be rejected. In addition, a significant proportion (again, up to 40 per cent) had only a marginal effect, with an unforeseen and possibly

negative impact. This suggests that only 20 per cent of systems have a positive effect on organizations.

OTHER INDICATIONS

Another indication is to study the rate at which the proportion of the organizational software budget consumed by maintenance has risen from 40 per cent in 1973 (quoted in Boehm, 1975), through 70 per cent in 1984 (King, 1984) to an estimated 90 per cent in 1990 (Oakley, 1988). This means that in many installations there may soon be no resources available to build new systems. It is suspected, although there is no direct evidence, that this maintenance backlog is chiefly due to changes to installed systems that do not meet user requirements.

A recent analysis (Humphrey, 1988) by the Software Engineering Institute, based at Carnegie-Mellon University in the United States, evaluated approximately 150 of the leading software producers in the United States. Five levels of software maturity of the development process used by the organizations were defined, where level 1 (the lowest level) was considered a chaotic process, with problem areas of progress planning and change control, and level 2 was a repeatable process with rigorous controls, but with problems in the design/code inspection and software process training areas; 12 per cent of organizations were on level 2, with 86 per cent of organizations on level 1.

CAUSES OF PROBLEMS

The main information systems problems and their causes, discussed in previous chapters, are summarized below. Table 12.1 shows why some problems are still outstanding, as the methods (including tools and techniques) that we have been considering only address some of the problems.

Table 12.1 Quality and productivity problems addressed by hard approach methods

Problem	Solution provided
Q1	—
Q2	—
Q3	Methods, tools
Q4	—
P1	Prototyping, group sessions
P2	—
P3	—
P4	—

Quality

1. *Wrong problem.* The wrong activities to assist are chosen, as the problem is not defined correctly or the system may conflict with organizational aims or strategies.
2. *Neglect of wider organization.* Wider social or psychological factors may be neglected, such as the degree of decentralization or centralization of the organization, or the factors of acceptability and usability.
3. *Incorrect analysis.* The right activities are identified, but errors may be made in analysing information needs due to poor development techniques.
4. *Wrong reasons.* Technology push or political pull.

Productivity

Productivity problems are caused by changing requirements, due to:

1. *Users change their minds.* Users refine their requirements as the project progresses, or there may be conflict between users.
2. *External events may occur that change the requirements.* Changes in external factors such as technology, legislation, the market or the political environment often change requirements.
3. *Implementation may not be feasible.* There may be implementation implications contained in requirements that are not feasible and that are recognized only during implementation and testing.

Another cause is:

4. *Poor project control.* Inadequate project resource estimation and tracking techniques.

Problem solutions

Two of the most significant problems are the first two quality problems. Methods that do not address these problems are characteristic of the hard approach to systems development, a typical example being the linear model discussed in Chapter 6. In contrast, methods belonging to the soft approach focus on these problems.

HARD APPROACH SOLUTIONS

The first assumption made by the hard approach is that the problem to be solved is logically based and has a computer solution. This assumption thus limits the range of problems that can be solved to those that possess a mathematical or logical solution. The second assumption is that the computer solution may be placed in the organization without taking account of the wider social or psychological factors with which the system will interact.

Many methods emphasize the *functional computer system*; that is, they concentrate on the tasks of a technical system, expressible in such forms as flowcharts, data flows and so on. The hard approach, then, neglects the important issues of defining the right problem and wider organizational factors. Requirements are thus drawn too narrowly.

SOFT APPROACH SOLUTIONS

There are two basic arguments why methods should take into account the wider impact of information systems on organizations. Firstly, the impact may be negative, as has occurred with methods based on the narrow technical view (Bjorn-Andersen, Eason and Robey, 1986), and, secondly, the full exploitation of computer technology may require a radical restructuring of the work process (Child, 1987).

With regard to the first quality problem, the soft approach allows for problem situations that may be investigated by a variety of problem definition techniques and emphasizes the determination of key organizational activities and strategies. A computer system will not necessarily result from this approach.

For the second problem, the soft approach focuses on wider issues that may be concerned with, for example, changing organizational structure or being concerned with employee job satisfaction. It also allows for the factors of usability and acceptability, which are perhaps equally as important as functionality, although less explicit, and which should also be taken into account when determining quality. These may be described as follows:

1. *Usability*. The intended users, with their inherent skills and capabilities, should be able to work with the system and use it as intended.
2. *Acceptability*. The minimum requirement for this is that the system should present itself to users in such a way that it does not threaten aspects of users' work held to be important. Ideally it will be perceived as being an active agent in assisting users to accomplish their desired work-related goals. It is often found that as many information systems bring about considerable change in organizational working practices, there is resistance to change. Although new working conditions may appear perfectly acceptable when viewed objectively, nevertheless to the existing employees, who may be cautious about change, they may represent a threat.

An issue that should be considered is: do the users want the system? It may often be the case that users have the system imposed from above (for example, head office) or outside (for example, a takeover) or they do not want it, as it may, in their view, lead to inefficiencies or redundancies.

An example of where psychological factors were not taken into account is described in Harrington (1991) where a manufacturing firm found that

Management/Department of Trade and Industry, London.

King, D. (1984) *Current Practices in Software Development: A Guide to Successful Systems*, Prentice-Hall, Englewood Cliffs, NJ.

KPMG (1990) *Runaway Computer Systems — A Business Issue for the 1990s*, October. Available from KPMG Peat Marwick McLintock, 1 Puddle Dock, Blackfriars, London, EC4V 3PD.

McFarlan, F. W. (1981) 'Portfolio approach to information systems', *Harvard Business Review*, vol. 48, no. 3, pp. 142–159.

Mantei, M. M. and T. J. Teorey (1989) 'Incorporating behavioral techniques into the systems development life cycle', *MIS Quarterly*, vol. 13, no. 3, pp. 256–273.

Mumford, E. (1981) 'Participative systems design: structure and method', *Systems, Objectives, Solutions*, vol. 1, no. 1, pp. 5–19.

Mumford, E. (1983) *Designing Human Systems for New Technology: The ETHICS Method*, Manchester Business School, Manchester.

Naur, P. and B. Randell (eds) (1968) *Software Engineering*, Report of a conference sponsored by the NATO Science Committee, Garmisch, Germany.

NEDO (1983) *The Impact of Advanced Information Systems*, National Economic Development Office, London.

Oakley, B. (1988) 'The issues of today' in *Software Engineering (Proceedings of Software Tools 88 Conference, London)*, Blenheim Online Ltd., London, pp. 1–4.

Tait, P. and I. Vessey (1988) 'The effect of user involvement on system success: a contingency approach', *MIS Quarterly*, vol. 12, no. 1, pp. 91–107.

Winfield, I. (1991) *Organisations and Information Technology: Systems, Power and Job Design*, Blackwell Scientific Publications, Oxford.

Wood-Harper, A. T. and G. Fitzgerald (1982) 'A taxonomy of current approaches to systems analysis', *The Computer Journal*, vol. 25, no. 1, pp. 12–16.

Wood-Harper, A. T. and D. J. Flynn (1983) 'Action learning for teaching information systems', *The Computer Journal*, vol. 26, no. 1, pp. 79–82.

Wroe, B. (1985) 'Towards the successful design and implementation of computer based management information systems in small companies' in *People and Computers: Designing for Usability*, M. D. Harrison and A. F. Monk (eds), Cambridge University Press, Cambridge.

13

Organization theory and information systems

Introduction

PSYCHOLOGICAL FACTORS AND INFORMATION SYSTEMS

The soft approach described in the previous chapter is based on the belief that system success should not be determined only in terms of whether the system performs the task it should, that is, in terms of functional or technical factors. In contrast, it emphasizes that the effects information systems have on *psychological factors* in the organization, such as job satisfaction, system usability and user acceptability, are also important for determining success. For example, jobs should be designed that maximize variable factors such as job satisfaction. Alternatively, for fixed factors, such as naïve or expert types of user, human-computer interaction styles may be designed to match. Work has also been carried out into matching other personal factors of individuals and different aspects of information systems. Factors studied include cognitive style, Jungian personality factors and leadership styles.

Another aspect of this psychological emphasis is the belief that increased user participation, particularly for obtaining user requirements, will in turn increase user acceptability, as users are given the system that they want. An information system should thus be designed to match the personal preferences of users, which will determine the types of information required, as well as the way in which information is acquired, stored and retrieved.

However, the system that users want may not necessarily be the best for the organization. This may be due simply to economic factors, as the desired system may be too costly, user bias (Moynihan, 1989) or the fact that a competent user may be an incompetent analyst (Avison and Wood-Harper, 1991). However, it may also be due to wider *social factors*, which are at the centre of this chapter.

- *Craft* (1). The supervisory level has high discretion and power, with co-ordination achieved through feedback. The technical level is weak, depending on reports from supervisors, and has little power or discretion.
- *Non routine* (2). Discretion and power is high for both groups, co-ordination is through feedback rather than by planning and group interdependence is high, meaning that both levels work closely together, as there is not enough information about production to enable the technical level to control the supervisors.
- *Engineering* (3). Great discretion exists in choosing tasks and planning is all important, with little interdependence between the two groups. Coordination is achieved by feedback of information for problem solving. Discretion and power by the supervisors is minimal.
- *Routine* (4). In these, the technical group has power over the supervision level, as they plan the work. There is low discretion for both, as problems do not generally occur, and planning is the co-ordinating force, as events can be foreseen. This approaches the model of a typical bureaucracy.

SUMMARY

Perrow's work is useful as it refines Woodward's one-dimensional scale of technological predictability into two dimensions. He also differentiates between two types of role, technical and supervision, showing how, for each of the four technology types, their activities vary. Implications for information systems will be discussed at the end of the chapter.

Accounting systems research

Management accounting systems have traditionally provided many of the control systems in organizations, and we will briefly describe work that has been undertaken in this area which has a direct bearing on the type of information systems we design.

CHARACTERISTICS OF ACCOUNTING SYSTEMS

The first group of authors, in a mixture of theoretical and empirical research, relate key characteristics of management accounting systems (dependent variables) to external, independent variables such as environmental uncertainty, and organizational variables such as technology, size, degree of decentralization and performance. The emphasis is on the use of such systems for controlling the organization's performance. As we might expect, they suggest contingent relationships between the independent and the dependent variables.

This work is relevant to managers planning information systems and designers engaged in strategic planning or analysis for a particular information system. The important question, to which this work gives a partial answer, is: what are the characteristics of a management accounting system that will best fit with the environmental and organizational factors in this situation?

We can identify three broad characteristics of management accounting systems:

1. *Complexity.* A fairly straightforward measure of complexity is used, where high complexity reflects a large number of different types of accounting system in existence or a large number of different accounting variables measured.
2. *Control detail.* This relates to the amount of discretion allowed to middle/lower management. Highly detailed controls measure through-put, while less detailed controls only set budget levels or outputs.
3. *Type of use.* The factor considered here is that systems may be used or they may be ignored, for various reasons.

Complexity

Work by Khandwalla (1972, 1977) found, in empirical research, that complexity was linearly related to organizational size, technological sophistica-tion, degree of decentralization and competitive environment. For high-performing organizations, as the degree of decentralization increases per-sonal controls are replaced by impersonal accounting systems, for coordina-tion and control purposes. Another relationship concerned the unpredictabil-ity of the environment, where a curvilinear relationship with complexity occurs. In predictable environments, personal controls are used, being replaced by impersonal controls to try to contain increasing unpredictability. However, there comes a point where controls are ineffective and manage-ment must respond to situations as they arise.

Another interesting result concerned the type of competitive environment. Where there was product competition, complexity increased, but for market-ing and price competition there was little and no association respectively with the degree of complexity.

Control detail

As organizations increase in size (Merchant, 1981) and as they decentralize, managers participate more in control setting, for example, in determining their budgets. In addition budgets are also less detailed, allowing managers more discretion in organizing their departments. As the unpredictability of the environment increases (Waterhouse and Tiessen, 1978), managers need more discretion to cope with this unpredictability, so controls are restricted to outputs rather than throughputs.

14
Strategic planning

Introduction

In this chapter we shall discuss the concept of strategic planning in some detail. The essence of the concept is that it is desirable to plan ahead for the introduction of information systems into organizations and that planning should take place on a strategic organizational level. A major motivation for planning is to ensure that information systems and their supporting information technology should fit in with organizational objectives, and not vice versa.

Strategic planning results in a strategic plan, consisting of two main elements, which are descriptions of:

1. Desired organizational activities and their information requirements.
2. Information systems, technology and resources necessary to support the activities.

Most authors state that it is important to separate the first element, which addresses organizational needs, from the second element, which addresses technological solutions. Although this is possible in the presentation of the plan, it is not so easy to separate these elements so clearly when the plan is being constructed, as some activities, for example electronic data interchange (EDI), only become possible through the use of certain technology.

Different approaches to strategic planning exist, and we shall examine three of these, the SISP approach (Ward, Griffiths and Whitmore, 1990), which we shall also illustrate with a case study, Information Engineering (Finkelstein, 1989) and the Dickson and Wetherbe approach (Dickson and Wetherbe, 1985).

Like the previous chapter, the intention is to extend the notion of the soft approach by focusing on one of the quality problems of systems development. The topic of strategic planning directly addresses the first quality problem, as it is concerned with defining the correct activities for the proposed system to assist and ensuring that the system will fit in with organizational strategy. Increasingly, it is being recognized that it is easier to do this in a top-down way, as all the proposed information systems can be compared and evaluated at the same time as organizational strategy is set.

General reasons for planning

In general, the main reasons for planning are to:

1. Decide on activities that will meet objectives.
2. Assign priorities to activities and allocate resources efficiently.
3. Anticipate the effects of the activities.

The planning process is important in organizations, as it includes the process of achieving consensus between individuals and departments, all of whom may have conflicting opinions as to important activities, priorities, resource allocations or even objectives.

REQUIRED ACTIVITIES

By deciding on the activities we want to do we also seek to avoid irrelevant activities or those that would have effects contrary to our objectives. We are also able to plan how those activities fit together. For example, on an individual level, at exam time, we might make a revision plan consisting of revision activities for the examination subjects. The plan would also include rest activities. By doing so, we try to ensure that we have covered all the necessary subject revision activities. We should resist any tempting activities that might arise during the revision period that are not on the plan, as they might detract from revision and rest time.

Similarly, the existence of a strategic plan for future information systems will only allow the development of systems that are in the plan. This seeks to avoid the situation where decisions to develop systems are made under the pressure of immediate events, such as, for example, aggressive advocacy by a user department. Such circumstances might not allow enough time for the relevance of the system to the organization's objectives to be properly determined.

PRIORITIES AND RESOURCES

Having decided on our activities, we cannot do them all at once, so we must decide their order of importance and the sequence in which we should undertake them. We also need to allocate resources, such as time, money, people and materials, to each of these activities in an efficient way and to maximum effect.

Returning to our revision analogy, some subjects need to be given top priority if they are key subjects, and some subjects will need to be revised before their dependent subjects. Top priority subjects may also require more time resources than others. It is easier to make these types of decisions in a rational, planning, context, with the information available that is necessary to make decisions.

For information systems, available resources will not allow us to build all the required systems at once, as there will generally be, for example, an annual information systems budget within which we must work. Some systems in the plan will be required immediately, whereas others can wait; some systems will be dependent on others. The resources required for development, operation and maintenance need to be determined, avoiding, for example, unnecessary or conflicting hardware, software, people or development procedures. Deciding on information systems in the absence of a plan may lead to biased prioritization or wasteful resource allocation. Planning may allow resources consumed by several systems, such as data-bases, to be shared.

ANTICIPATE EFFECTS

If we know what our activities will be, it may be possible to predict, to some extent, the effects of the activities and their effects on the environment. When you consider your revision plan, you might realize that for the next two or three months you are likely to be a rather moody person, affecting the friends with whom you live. Similarly, a planner might predict that the side-effects of planned information systems on workforce attitudes towards the organization might be negative, unless, for example, a policy of education and training in information system use is adopted.

STRATEGIC PLANNING — PROBLEMS AND SOLUTIONS

Strategic planning implicitly addresses three problems.

Quality

Many information systems are delivered that are never used, as they do not meet organizational requirements (Fairbairn, 1989). Systems may not fit in with organizational strategy or may not address the most relevant activities for the organization.

Productivity

Many information systems are delivered late and over budget (Fawthrop, 1990). The problem of poor productivity is occurring in a context where systems are having an increasingly critical impact on organizations. Early data processing systems affected only a localized part of the organization, whereas systems may now link many areas or provide an organizational infrastructure. A problem with the timing of system delivery may thus seriously affect the whole organization, making decisions concerning systems more risky than before.

Technological effectiveness

New technology can improve organizational performance, but it is difficult to determine the best way in which it may be used. Imagination and creative thinking are required to apply new technology to an organization's activities.

Solutions

For the first problem, strategic planning aims to plan for information systems that fit in with the objectives and activities of the organization. The second problem is addressed by rationally allocating resources in advance, planning an infrastructure that takes into account evolving technology and producing global information requirements that will act as a basis for future application development. For the third problem, authors do not claim to do other than focus attention on the issue.

Definitions and objectives

Strategic IS/IT planning is taken to mean planning for the effective long-term management and optimal impact of information, Information Systems (IS), and information technology (IT), incorporating all forms of manual systems, computers and telecommunications. It also includes organisational aspects of the management of IS/IT throughout the business.

(from *Strategic Planning for Organisations* by J. Ward, P. Griffiths and P. Whitmore, Wiley, Chichester, 1990)

We will use the terms *corporate planning* or *strategic planning* for the management focus on planning, and *systems planning* for the DP emphasis. As an organization becomes more reliant on computers, its data processing hardware and software resources need to be closely aligned with its corporate plan.

(from *An Introduction to Information Engineering* by C. Finkelstein, Addison-Wesley, Wokingham, 1989)

Four objectives are noted by Ward, Griffiths and Whitmore, which are:

1. Alignment of IS/IT with the business, and the impartial determination of priorities for development.
2. Competitive advantage through IS/IT, by exploiting opportunities and combating threats in the external business environment, using the strengths of the organization.
3. Building a rationalized and flexible platform for the future.
4. Improved budgeting and resourcing and an ability to develop cost/benefit cases for long-term or infrastructure developments, or those where a payback is difficult to define.

It may be seen from the above definitions and objectives that strategic planning may be rather a complex exercise, trying to establish (a) the correct organizational activities, (b) the information needs and the information systems required for those activities, (c) the efficient allocation of organizational resources to implement the activities, (d) a technological structure to support, in an efficient and integrated way, the planned information systems and their evolution and (e) competitive advantage through novel applications of new technology.

Information systems eras

To understand why recent developments in information systems have made planning more necessary for organizations, we shall briefly discuss the 'three-era' model of information systems (Ward, Griffiths and Whitmore, 1990).

In the first era, termed the data processing (DP) era, dating from the 1960s, the aim was increased efficiency, in terms of cost/input, and most information systems only automated the functional activities in organizations, including transaction handling and some exception reporting. The second era, the management information systems (MIS) era, which began in the 1970s, emphasized effectiveness, and was concerned with user-initiated enquiry systems and information analysis for assisting management activities. These management systems shared a common base of information with DP systems.

The strategic information systems (SIS) era, dating from the 1980s, is the current era, and it is characterized by new types of system, which, for example, incorporate more intelligence than before (expert systems, decision support systems). Another feature is that they may process a wider variety of information, such as graphics, image and voice, as well as traditional text. Their aim is more ambitious and is to improve competitiveness.

The planning approach characteristic of the DP era was to build systems bottom-up, in a piecemeal manner. Where the aim was to improve efficiency, this had some success, but where effectiveness or competitiveness are concerned, integrated systems are required, which cannot be developed in this way.

Ward, Griffiths and Whitmore argue that many organizations are now contemplating the possibility of developing strategic information systems that directly address the increasingly evolutionary nature of organizations and their objectives. Common types of strategic information system are:

1. Systems that link the organization's technology directly to important external organizations and individuals, such as customers and suppliers.
2. Systems that integrate and disseminate information in an organization over established barriers such as different roles or departments.
3. Systems that enable organizations to develop new products or services based on information.
4. Systems that provide top executives with information to assist strategic activities.

As such systems are likely to have a direct and immediate effect on the organization's competitive position and are likely to be at least partially determined by information system requirements from external customers and suppliers, the involvement of the most senior levels in the organization are required in their planning.

In addition, as the systems depend upon a wide variety of internal information that is well integrated, they require a good foundation of functional and management systems. As many different types of system will communicate, they must all be based upon compatible technology. This includes shared data (common data definitions and structures), software (compatible databases, operating systems) and hardware (open systems or industry-standard computers, telecommunications). Such integration and compatibility must be planned and managed. As pointed out earlier, such systems are inherently risky, as a system problem may have a potentially large effect on the organization.

Strategic planning process

There is no accepted procedure for strategic planning and there is no standard for the output — the strategic plan. This reflects both the immaturity of the area, as well as the fact that planning is generally linked to the organization's own planning procedures. We may, however, find aspects in common between different approaches, and a general model for the process is seen in Fig. 14.1.

PRE-PLANNING

Before the planning study itself begins, there is usually a stage termed pre-planning, as shown in Fig. 14.1. This begins by gaining a preliminary idea

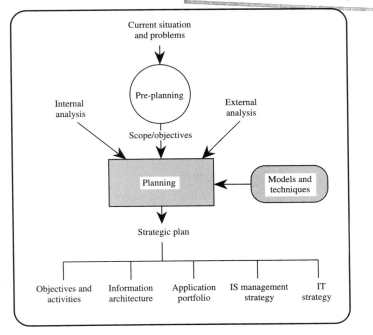

Figure 14.1 General model for strategic planning

of the organization and any current problems, and it sets the scope and objectives of the planning study.

Many organizations are so large (Olle *et al.*, 1988) that it is not generally possible to produce a single plan for the whole of the organization; therefore the required scope, which might be an autonomous operating company or a division, must be set. Problems or opportunities that are known may also be used to focus on a certain set of activities.

The objectives of the planning study are also set, which will emphasize different elements of the strategic plan. In addition, pre-planning will concern itself with normal project planning considerations such as resource allocation, defining report contents and deadlines, and team composition. A strategic plan is normally concerned with a span of about five years.

PLANNING

Input

After pre-planning has defined the initial parameters, the planning study itself can begin. Figure 14.1 shows two different types of input. In internal analysis, the organization is examined with a view to determining its objectives, its current activities, whether current activities are meeting objectives, its problems and possible problem solutions. This analysis is concerned solely with business or organizational issues. Another level of analysis considers information needs and how they are being met by current information systems. The skills and attitudes of organizational employees are also determined.

External analysis examines significant external entities, such as suppliers, customers, governments and markets, to determine any problems or opportunities the organization may be faced with. It also studies the products or services of the organization in relation to the market and competitors. Information systems belonging to comparable organizations, as well as technical trends, may also be considered.

Output

The output of planning is the strategic plan. The contents and emphasis of this vary between different approaches, but the main elements include descriptions of the required:

1.   Organizational objectives and activities
2.   Information architecture
3.   Application portfolio
4.   Information systems management strategy
5.   Information technology strategy

New or refined objectives (1) may arise from the planning study, and these may lead to new activities being defined to meet the objectives. The information architecture (2) is a high-level definition of the entities, activities and their relationships which are the basic components of the information needs of the activities. It is used as a basis for IS development and usually emphasizes entities common to many systems.

The term *application portfolio* (3) is used to refer to the set of information systems applications in an organization; this element contains descriptions of the information systems and their relative priorities, which are required to support the information needs of the organizational activities. The information systems management strategy (4) is a detailed plan for various aspects of the management of the application portfolio and information architecture, including resource considerations, hardware vendor specification and organization structure. The IT strategy (5) describes the procedures and technology that will be used, including factors such as networking, hardware and software standards, systems development and project management approaches and employee training.

Models and techniques

To transform the inputs into the outputs of the strategic plan, various techniques and models are used. For example, there may be models for analysing the application portfolio of the organization, competitive forces in the market or information needs of activities.

SISP approach

The main features of the SISP (Strategic Information Systems Planning) approach (Ward, Griffiths and Whitmore, 1990) may be seen in Fig. 14.2. The pre-planning stage is termed planning for planning and the outputs of the planning stage consist of the business IS strategy, containing the required information systems in the application portfolio, the information technology (IT) strategy for implementing the portfolio and the IS/IT management strategy, which will monitor and control implementation of the plan.

PLANNING FOR PLANNING

The aim of this stage is to set the terms of reference for the planning study proper which will follow. It consists of six main steps, shown in Fig. 14.3.

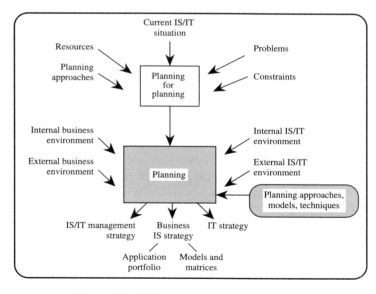

Figure 14.2 SISP approach to strategic planning

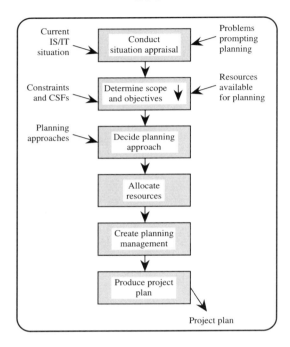

Figure 14.3   SISP steps in planning for planning

Conduct situation appraisal

Firstly, a preliminary analysis of the current role of information systems in the organization is made. This might, for example, determine any applications backlog or look at user satisfaction. Secondly, it obtains a description of the stimuli or problems that have prompted the need for a planning study, for example a desire to explore new technology for bringing about new products or markets, or major problems facing management that require a solution.

Determine scope and objectives

This sets key features of the planning study such as:

1. *Scope.* This involves deciding which part of the organization will be studied and setting a time limit to the study. For large organizations, the level of the business unit has been found to be a satisfactory level, while setting the scope at the corporate level has been found to be too wide. These levels are shown in Fig. 14.4.

    A business unit is part of an organization that, in the commercial field, for example, sells a group of related products or services to a specific set

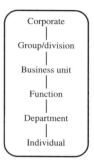

Figure 14.4 Levels within an organization

of customers and competes with a well-defined set of competitors. It will be assisted by functions such as accounting and marketing.
2. *Objectives.* These may vary, as, for example, an organization may require an emphasis on integrated information, a resource allocation plan for already agreed information systems or new, strategic, systems.
3. *Constraints.* There may be specific constraints placed on the study at the outset, stating what the study should not do. For example, it should not study certain parts of the organization or recommend increased IS/IT budgets.
4. *Critical success factors.* These are the central issues which will be used to judge the success of the planning study, and they may reveal a 'hidden agenda' behind the terms of reference. They are the key factors essential for meeting organizational objectives.
5. *Deliverables.* Typical outputs from the planning study would be: main sections of final report, report dissemination procedures and any 'soft' outputs such as training of organization staff.

Decide planning approach

The authors state that there is no one best way of conducting the planning study, and the characteristics of the organization will need to be taken into consideration. They also suggest that the responsibility for producing a successful outcome depends heavily on the project leader. Decisions are required concerning, for example, the techniques to be used for analysing the business and the models to be used for analysing information needs.

Allocate resources

The team to conduct the planning study must be balanced between user and IS representatives. Seniority of the team members should be very high, as far-reaching decisions may have to be made and implemented in the organization. Automated diagramming tools are useful for documenting information requirements.

Create planning management

An important figure is the management sponsor, who ideally is a senior executive or director of the organization who assures management commitment to the planning study. He or she will generally chair a steering committee, which will meet periodically to review the deliverables of the project team, as well as reporting project progress and results to the board of directors. Individuals from the organization who will participate in the planning study interviews, workshops or discussions should also be identified at this stage.

Produce project plan

The different stages of the planning study need to be identified, with their associated techniques and methods, emphasizing checkpoints, deliverables, project time and relative time of each of the stages. Planning studies generally range from three months to a maximum of six months. A critical factor affecting the length of the study is the number of individuals to be interviewed. The project plan should also contain a summary of the costs and benefits of undertaking the proposed study. These should directly address any problems in the organization.

Summary

This pre-planning process is thorough and provides a lot of detail which will be required later in the planning study. It emphasizes the need to decide the scope of the planning study, suggesting the business unit as the best level of study, as well as focusing on the current IS/IT situation and determining any problems.

PLANNING

Internal business environment

The aim here is to (a) thoroughly understand the objectives of the organization, (b) ascertain current activities and determine whether these are meeting the objectives, (c) determine required activities, (d) decide on the information needs of required activities.

- *Organizational modelling.* This step is an initial fact-finding step, gathering information for later use. The type of information gathered relates to (a) organizational factors such as job descriptions, employee skills, management styles, trade union representation, (b) the level of use of technology internally and in competitors, (c) the external environment, including competitors and legislation, (d) the influencers in the organization, (e) key processes that produce products or services. An important use of this task is to ascertain the 'culture' of an organization, so that changes that result from the strategic plan are likely to survive.

- *Analyse business strategy.* Organizations have objectives and strategies for achieving those objectives, which are important as they determine the general framework for deciding what the organization should do. The concepts generally used when discussing objectives are:
  - *Mission.* The highest level statement of the long-term aims. An example is the mission of the Manchester Training and Education Council, which is:

    'to play a leading role in the creation of a successful and growing economy in the Manchester area by:
    · encouraging employers to develop a workforce with the skills, knowledge and competence to make them competitive in world markets;
    · giving every individual access to the training which matches their potential and aspirations with job opportunities;
    · stimulating and supporting business growth and enterprise.'

  - *Objectives.* High-level targets, more specific than the mission, which are usually quantified, possibly with deadlines, and which may be at global or business unit level. For example, some objectives of a business might be:
    (a) General, to increase turnover annually by 5 per cent.
    (b) Customer service, to reduce complaints by 50 per cent.
    (c) Services division, to gain 20 per cent of markets y and z.
  - *Strategies or tactics.* These describe how the objectives will be achieved. For example, for objective (b) above, the organization might increase the effectiveness of the customer service department or improve the quality of its products. The Manchester TEC has six strategic aims, two of which are:

    'To improve the competitiveness of business by developing business management skills and gaining recognition of the contribution education and training can make to successful business planning.
    To determine and to disseminate more quickly and effectively, information and advice on present and future needs and resources in the local economy.'

- *Analyse critical success factors.* Critical success factors (CSFs) are defined as the 'few key areas where things must go right for the business to flourish'. They are typically short statements of what are perceived to be crucial organizational variables, distinguished from the mass of variables that have a bearing on organizational performance. For example, in a business, a CSF might be the level of advertising, and in a school, it might be the quality of teaching.

CSFs can be used to interpret mission and objectives and, as they describe the areas that must turn out correctly, can also be used to help in deciding future activities, if necessary.

- *Analyse business activities.* These consist of operational activities, which are activities performed by the organization intended to meet the objectives via the strategies, and performance measurement activities, which measure how well an activity is supporting the achievement of an objective.
- *Analyse information needs.* This step analyses the above activities to a greater level of detail to determine their information needs, that is the information which they require to operate effectively. A central aim of the SISP approach is to shape these needs into an *information architecture*, which is to be used as the basis for application planning. For the Manchester TEC, for example, the information needed to support the second strategy described would be analysed here.

Several types of model, such as entity models, activity decomposition diagrams, data flow diagrams and various matrices are used. Entity models will show key entities only, which may contain lower level entities. Activity decomposition diagrams decompose the business activities to two or three levels of detail and data flow diagrams show activities and the data they use.

The activity entity matrix shows the entities that are used by each business activity. Clusterings of entities and activities may be used to create an architecture matrix, showing proposed databases and application areas.

Models may exist for the current situation, a future 'ideal' situation and different transition points towards the ideal. An aim is to identify redundant, inconsistent or inappropriate activities and data in the models showing the current situation, removing them from the models for the future information architecture.

External business environment

The study of the external business environment considers three broad areas: (a) external environment, (b) pressure groups and stakeholders and (c) business planning processes. The SWOT (strengths, weaknesses, opportunities, threats) framework is used to analyse the information, and then creative thinking and brainstorming sessions are used to balance these four basic factors.

- *External environment.* There are many factors in the external environment that may influence organizations. Some important factors are:

  — *Economy.* Tax rates, inflation, currency strengths.
  — *Law.* Legislation may particularly affect public organizations.
  — *Technology.* Rapid technological change is becoming more evident, affecting telecommunications and robotics, to take two examples.

- *Pressure groups and stakeholders.* Pressure groups are individuals or organizations who make demands on the enterprise and who expect a rapid response to that pressure. Stakeholders have a stake in the organization and expect to receive benefits. The importance of the analysis is that threats should be identified so that the organization may defend itself, but, in addition, opportunities may be posed that can be exploited.

  Pressure groups include:

  — *Shareholders.* Groups such as pension funds, who possess a significant proportion of an organization's shares, can affect the share price by buying or selling large quantities of shares, or can press for alterations in management.
  — *Competitors, suppliers, customers.* These groups can exert direct pressure by, for example, demanding rapid payment or withholding payment.
  — *Employees.* Employees of the organization may exert pressure in areas such as equal rights for women.

  Stakeholders include:

  — *Shareholders.* These will expect dividends to increase, as well as the value of their shares.
  — *Government.* Increases of tax revenues, jobs, investment in research and development will be factors in which the government is interested.
  — *Public.* Increasingly, commercial organizations are playing a larger part in the community, sponsoring educational or artistic associations, or creating community centres in the local towns.

- *Business planning.* First of all, the organization is analysed for strengths and weaknesses, including financial resources, people and their skills, plant and equipment, research and development status, and the products or services being sold. The next task involves an analysis of competitors so that the position of the organization in the market may be clearly seen. Among the factors considered are:

  — *Markets, market segments and market share.* The demand (the market) that exists for the products or services (for example, newspapers) is analysed, as well as parts of the market, for example, quality newspapers (market segment). Current and potential competitors are analysed for threats and opportunities.

— *Product life cycles.* Each product has its position determined within the basic product life cycle. For example, a product may be a promising entry from research and development, or it may be becoming obsolete.

— *Competitors.* Analysis of current and potential competitors is made for strengths, weaknesses and their products or services.

— *Future competitive action.* The possibility of new products or services being developed by competitors to exploit existing or new markets is considered.

Internal IS/IT environment

- *Current application portfolio.* The current information systems will be analysed, using, for example, the three eras model discussed above, to understand their value to and impact on the organization and its objectives.
- *Current IS/IT environment.* There are several factors concerning the current use and management of IS/IT within the organization that may have an impact on future use. These include IS organization, hardware and software assets, investment and expenditure, methods and procedures for tasks such as systems development, quality control and project management, and end-user computing.

External IS/IT environment

The external IS/IT environment is analysed to gain awareness of technical trends and opportunities for using IT in an innovative way. This can be achieved by studying competitors or other comparable organizations.

OUTPUTS

SISP produces three main hard outputs, in addtion to soft outputs. Soft outputs are the skills, levels of awareness and motivation that may be produced by the planning study. The three hard outputs are (a) business IS strategy, (b) IS/IT management strategy and (c) IT strategy.

The relationship between these three components is that the business IS strategy is a statement of 'demand' in terms of the organizational requirement for information, systems and technology, the IT strategy states the 'supply' for this 'demand' and the IS/IT management strategy contains policies for satisfying and balancing demand and supply.

Business IS strategy

- *IS strategy and policies.* These state how the organization will use IS/IT to help it achieve its objectives. They cover business strategy (internal and

- *Key business areas.* These are the financial, land and property, community and internal administration areas; some systems will belong to more than one area.
- *Critical success factors.* These are: to meet statutory demands (that is, changes in the law); to maintain effective financial systems; to support existing systems; to transfer control of systems to customers (that is, divisions); and to reduce the level of manual administration.

Application portfolio

There are 103 systems to be considered, not including 'personal' systems, and the inventory of these systems is: current — 38, under development — 44, proposed — 21. However, of these, only 17 are identified as being priority systems, and these are shown in Table 14.1, by division. Figures indicate how many information systems belonging to that division are not prioritized. Systems shown under the 'central' division are those where more than one division depends on the output of that system for successful operation.

The personnel system will achieve a common personnel database for all departments, as there are currently three systems in use. Office information systems aim to extend the automation of manual processes into areas such as word processing, diaries, spreadsheets and electronic mail. The geographical system is intended to maintain a database of maps and associated information relating to property and land owned by the authority.

Information architecture

An important element of the plan concerns authority data, where the aims are to reduce redundancy and to improve sharing, which is part of the overall plan objective. Redundancy has occurred mainly for technological reasons, where past systems have not been able to share data, and the benefits of eliminating redundancy are reduction in storage space and input, as well as reduction of errors.

Decisions relating to information systems, the rationalization of data and the creation of databases are as follows. In the financial area, improvements to central financial systems will gradually remove the need for local systems that duplicate data. The new geographical system will provide a focus for the integration and access of various data relating to land and property, in conjunction with a digitized ordnance survey map. Data cleaning activities are required here. Several divisions, such as housing and environment, hold relevant data in manual or computer databases which may be integrated with the geographic data. For the social services division, only manual records exist, which will be cleaned and built into a set of local databases, and eventually integrated for the two new applications (shown in Table 14.1); other divisions such as education and housing also have relevant data.

Models, matrices and tables

Table 14.2 shows a matrix of critical success factors with prioritized information systems, indicating the factors that were instrumental in allocating priority to each of the systems.

IT STRATEGY

Information resource management

The central element in the management of data is to be a centralized data dictionary, which will define details of all data items, referring to data owner, input point, sensitivity and associated systems. Such a definition will eventually reduce redundancy and provide a basis for increased sharing.

Technical means

● *Introduction*. The main feature of the current technological situation is that the various information systems of the authority, in the absence of a single supplier policy, have evolved in different proprietary forms, leading to unconnected 'islands' of information. The overall aim is to integrate

Table 14.2   Critical success factors for each prioritized information system

Information systems	Support existing systems	Statutory demands	Financial systems	Office systems	Other systems
		CSFs			
Financial information	√		√		
Debtors	√		√		
Creditors	√		√		
Poll tax information	√	√	√		
Stores/purchasing	√		√		
Housing benefit	√	√	√		
Rates	√	√	√		
Payroll	√	√	√		
Superannuation	√	√	√		
Personnel					√
Direct labour control			√		
Geographical system					√
Stock repairs			√		
Office information systems				√	
Minors information system		√			√
Community care		√			√
Project costing			√		

these systems, thus allowing access to central or divisional data, and also to provide facilities for the sharing of local processing resources. This will be achieved by a move in the direction of open systems standards and the adoption of the client/server approach (a distributed and shared processing approach).

- *Communications systems/networks.* Networking plays a large part in the overall plan, as it provides the basic vehicle for integrating together different systems. This will be done by a single data network, connecting all sites and processors, with the aim of allowing single terminal/PC access to all relevant applications over all host systems.

  Until OSI (open systems interconnection — the emerging international standard for connecting different types of hardware and software) is fully implemented, industry standards such as Novell NetWare will be used for LANs, with the Authority WAN being based on extended-LAN topology with bridging and routing technology. TCP/IP will be the standard for terminal attachment and file transfer between different manufacturers.

- *Hardware.* Current hardware should be used as much as possible to protect existing investment. Resources do not allow the authority-wide replacement of dumb terminals by PCs or intelligent workstations, as factors such as training, administration and development have to be considered as costs, although the cost gap is diminishing. However, where possible, PCs should be installed instead of terminals, networked as clusters to share resources such as data, printers and disks.

  PC standards are: colour VGA monitor, 3.5 inch 1.44 M floppy disk, hard disk capacity of at least 40 M. Diskless configurations should be available. Memory should be a minimum of 640 K with expansion to 4 M. Software supported must be: MS-DOS, Windows 386, OS/2 or Unix V/386, depending on configuration.

  Workstation standards are: colour VGA monitor or high-resolution graphics screen, storage capacity up to 300 M and 1.44 M floppy disk drive. Tape back-up capacity should be available. Memory should be a minimum of 2 M with expansion to 8 M.

- *Software.* For operating systems, the authority currently uses five different types, which will be reduced to three: open VME, Unix and MS-DOS (corresponding to central, divisional and personal applications), based mainly on intent for conformance with open standards (open VME, Unix) and the *de facto* industry standard MS-DOS. Cases that require different systems should be clearly cost/benefit justified.

  For applications software, a set of selection criteria for software packages has been established, as well as a recommended list. These include, for central applications, Querymaster and Reportmaster; for divisional applications, SQL and Officepower; and for personal use, Lotus Symphony, Quattro Pro, Wordperfect, DBASE 4 and Pagemaker.

  Applications packages should be used in preference to bespoke software, although PC users may develop applications using spreadsheets and databases. Customization of packages will in future be limited to interfacing, report production and 'cosmetic' changes. All applications software must run under one or more of the three operating systems chosen.

Management issues

- *Application management.* The PRINCE project management method will be introduced with the aims of improving systems quality and involving users more closely in systems development. Systems development will be undertaken using the project concept, involving a set of specific objectives and a set of identifiable stages. Quality assurance procedures will be developed and applied to systems development, in conjunction with PRINCE. This will be implemented in stages until 1993. Project managers will be 'customers' (that is, divisional members), with technical project leadership being provided by either the divisional IT section or the central IS group (see IS/IT organization below). Another aim will be to increase end-user computing by increasing systems where users can provide input and use fourth generation languages.
- *Organization and resourcing.* The roles of analyst and programmer will be combined to form one role of analyst/programmer. Training consists of general training, where spreadsheets, database and general PC apprecia-tion are currently taught, which will be extended to all the software packages recommended. Specific training concerns new systems that are developed for the divisions.

IS/IT MANAGEMENT STRATEGY

The general aims here are to control and optimize resource allocation, to emphasize quality assurance and to establish frameworks allowing greater customer control and involvement in future projects.

IS/IT organization

A central IS group, consisting of approximately 110 staff, will have the following functions: training, systems development, quality assurance and computer centre. Each division has an IT section, responsible for meeting the divisional IT needs. A senior manager, not necessarily a member of the IT section, is responsible for IT in each division, who advises the divisional director on IT matters. A central strategy group, chaired by the Chief Executive, plans and co-ordinates all IT development, and each division has a representative on this group, either the director or manager responsible for IT.

The members of IT sections have functional, but not line, responsibility to the central IS group, as this group has the responsibility of identifying divisional IT needs. The main responsibility of the central IS group will be to satisfy central authority IT needs, such as hardware and software procurement, network management, training, methods and standards, central systems development, operational services and consultancy. A data manager will be required for the central data dictionary.

IS accounting

Costs for services provided by the central IS group have been agreed for the current financial year, and costs for divisional IT section services will be introduced by next year. The basic framework of client–contractor agreements has been finalized and will be introduced to all new projects as the basic means of defining responsibilities and service costs and timetables.

CASE STUDY CONCLUSIONS

We may note that this is a plan for a three-year period, covering all of the authority, and hence the level of detail is fairly high, with no detailed information architecture, for example. The strategic objectives highlight those aspects of the current situation that need to be improved. In the application portfolio section, it is surprising how many systems belong to 'central' as opposed to the divisions, given the objective of devolving systems to customer ownership. Note that we can infer that a significant proportion of resources are committed to maintaining the current systems.

Turning to the IT strategy, the financial situation does not permit rationalization of the different hardware and software platforms currently possessed, so that reliance on emerging open standards based on networking sites and machines together is the option chosen to reduce data redundancy and increase sharing. There is also a recommendation to do as little bespoke development as possible, with an emphasis on end-user computing to use the authority data that should become more easily available.

Information Engineering

INTRODUCTION

The aim of *strategic planning* within the Information Engineering approach (Finkelstein, 1989) is to align the data processing hardware and software resources to the corporate plan. Finkelstein points out that effective strategic planning is difficult, as the impact of alternative business plans on IS plans is hard to determine. In addition, planning must always be an inherently flawed process, as alternative approaches are not always obvious and later problems may not be apparent due to an insufficient level of detail in the plans.

A solution to this problem is shown in Fig. 14.6 and rests upon a careful evaluation of the effects of the plan on the organization. In the figure, *strategic implementation* consists of a high-level 'implementation' of the plan, in the form of entity diagrams, and *strategic management* analyses this, making plan refinements if necessary. Also, the feedback process must give early results, and the whole planning process should be continuous, not stop–go.

The approach consists of two types of planning, *formal* and *informal strategic planning*. The output from planning is a set of strategic statements. These basically contain the business activities required for the organization, together with an indication of how their information needs may be determined.

FORMAL STRATEGIC PLANNING

The approach envisages *formal strategic planning* being adopted by an initiative from senior management. The steps may be seen in Fig. 14.7.

Identify current strategy

The aim of this step is to determine the current strategy of the organization, business unit or other part of the organization being studied. Among the

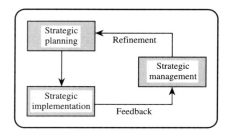

Figure 14.6 Strategic planning and its evaluation within Information Engineering

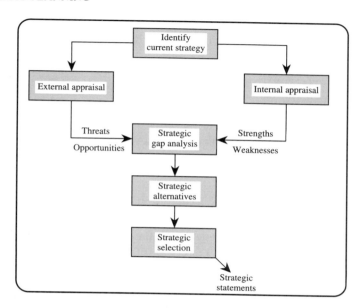

Figure 14.7 Steps in formal strategic planning

factors to be identified are the purpose and activities, as well as policies, issues, plans and budgets. Future directions are also considered as well as possible changes.

External appraisal

The external appraisal consists of an analysis of the organization's products or services with regard to environmental elements such as competitors, industry, markets and legislation. A life cycle model consisting of several stages between development and decline is used as the basic method of analysis. The following factors are considered:

- *Threats/constraints and opportunities*. Each product/service has its position in its market defined in terms of the life cycle, and environmental factors such as economic and legislative factors are used to identify threats (for businesses) or constraints (for public sector organizations). Opportunities are also identified.
- *Strengths and weaknesses*. The business unit is assessed in relation to the industry and the market, looking at potential for competition from other organizations, and strengths and weaknesses are evaluated.
- *Position of competitors*. Each competitor is then compared to the business unit, as above.

- *Effect of interaction.* The effects of various actions by the organization on its competitors (or political parties), and vice versa, are determined, together with likely responses.
- *Environmental change.* The effect of environmental change on the business unit, industry and competitors is evaluated for threats and opportunities.

The competitive position of the business unit is thus determined by this process, together with threats and opportunities.

Internal appraisal

The internal appraisal looks inside to determine the strengths and weaknesses that will respond to the external threats and opportunities.

- *Strengths and weaknesses.* These are determined for the business unit using factors such as plant productivity, financial profitability, cash flow, management effectiveness, and employee skills and motivation.
- *Distinctive competence.* Functions or areas are analysed for particular competence.
- *Comparative advantage.* The areas of competence are grouped to emphasize the characteristics that might provide an advantage over competitors.
- *Comparative disadvantage.* Areas that are weak are similarly grouped to emphasize comparative disadvantage. Remedies may be identified.
- *Resource sensitivity.* Strong areas are evaluated for the extent of their dependence on resource variations, for example, a car manufacturer with only one supplier of sheet steel or a government agency dependent on a particular political party.

The result is a set of performance indicators in terms of strong, average or weak for the business unit, indicating its relative advantage or disadvantage to competitors.

Strategic gap analysis

The effectiveness of the current strategy is now determined, in the light of the internal and external appraisals.

- *Specific comparative advantage.* This compares those areas of competence with their external competitive position, identifying specific comparative advantage.
- *Performance gaps.* The effect of the current strategy on threats and weaknesses is identified and likely performance gaps identified.
- *Feasible opportunities and potential difficulties.* Areas of comparative advantage and distinctive competence are compared with opportunities,

to determine feasible opportunities. Similarly, strengths and weaknesses are compared to identify potential difficulties.
● *Strategic agenda.* Performance gaps, as well as feasible opportunities and potential difficulties, form the agenda for the future.

Strategic alternatives

This step analyses the strategic agenda elements into three groups: performance gaps, potential difficulties and feasible opportunities, determining each element in terms of alternatives, directions, assumptions, alternative assumptions, threats or weaknesses and evaluation. The emphasis should be on strengths and opportunities, rather than on weaknesses or threats.

Strategic selection

Each of the above alternatives is evaluated in terms of: contribution to objectives, confidence in the procedure that generated the alternative, political emphasis, corporate or legislative constraint and relevant strategic emphasis such as proactive or reactive. Finally, the strategic statements are produced which describe the agreed objectives and activities.

Summary

Formal planning is an approach that analyses an organization and determines its strategic direction, in terms of the activities that reflect its objectives and make most use of the strengths inside and the opportunities outside the organization. However, Finkelstein comments that this may be a lengthy process, requiring management commitment and time, which may not be available initially. In this case, informal strategic planning may be used.

INFORMAL STRATEGIC PLANNING

There are two distinct stages in *informal strategic planning*, termed *strategic* and *tactical*; the steps in the strategic stage may be seen in Fig. 14.8.

Determine mission and purpose

Mission and purpose have a similar meaning to the concepts discussed in the SISP approach above. They describe the current aims of the organization or business unit. However, account should also be taken of the future, in terms of the direction in which the organization will and should evolve. The mission and purpose statements should not be longer than one page of text.

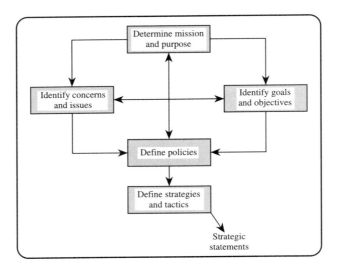

Figure 14.8   Steps in informal strategic planning (strategic stage)

Identify concerns and issues

These provide a focus for areas of threat or areas subject to weakness, for example dependency on outmoded technology or on products at the end of the life cycle. The concerns and issues may have prompted the planning process and may modify the stated mission and purpose.

Identify goals and objectives

Goals set long-term results for the organization, while objectives set short-term result targets. Goals and objectives are quantifiable and define what is to be achieved as well as the amount of achievement. These should obviously be derived from the mission and purpose, and should refer to a unit of performance, a level of performance and the time when the level should be achieved. For example, an objective might be to achieve £x profit within two years.

Define policies

Policies are qualitative statements, which provide a plan for future action and which help employees decide what is relevant and what is not.

Define strategies and tactics

Strategies define how objectives can be reached and tactics are refinements of strategies. A strategy to achieve the objective of £x profits above might involve developing a new product.

Summary

The resulting set of mission, purpose, goals, objectives, concerns, issues, policies, strategies and tactics comprises the strategic statements. In addition, Information Engineering lays down guidelines for translating these statements later into information needs, according to the following rules:

Rule 1. Policies and issues relate directly to entities.
Rule 2. Goals and objectives relate directly to attributes.
Rule 3. Strategies and tactics relate directly to associations.

INFORMAL STRATEGIC PLANNING (TACTICAL STAGE)

The second stage of informal strategic planning is shown in Fig. 14.9. These add more detail to the strategic stage as they define markets, products or services, and distribution channels for the organization. The justification for

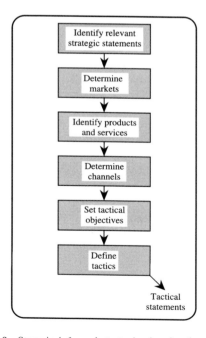

Figure 14.9   Steps in informal strategic planning (tactical stage)

the stage separation is that rapid change makes organizational environments more dynamic than in the past, necessitating decisions being made at lower levels in the organization than before. However, these decisions need the overall direction of the strategic statements.

Identify relevant strategic statements

The tactical area of the part of the organization being considered has those strategic statements selected that are relevant to the area.

Determine markets

This defines the customers of the organization, whether a population segment served by a government department or a specific industry served by a business. The specific market segments (internal or external) served currently by the business unit are described in detail, as well as possible future markets, due to factors such as changes in technology or legislation. A distribution warehouse is an internal market used to hold goods delivered by a production department. Detail includes market needs, who and where are the customers/ clients, what they will pay and what they consider as value.

Identify products and services

Following the definition and identification of markets and market needs, products or services are identified which address the market needs. Broad outlines of future products or services should be considered.

Determine channels

The channels of an organization concern the way in which products or services are delivered to customers. This delivery requires resources, in terms of people, money and equipment. Resources are required to make a product, perhaps assemble it using different departments, distribute it and sell it. The emphasis is procedural, concerning itself with turning inputs into outputs, for example detailing the passage of orders by telephone or writing from the orders department to sales, to production, to manufacturing and so on.

This step then focuses on how products and services are delivered currently and in the future, especially taking into account new technology.

Set tactical objectives

Tactical objectives are short-term performance indicators and are set to be measurable. They indicate the level of acceptable performance.

Define tactics

Tactics describe the activities that will achieve the tactical objectives of the organization.

Summary

The set of markets, market segments, products/services, channels, tactical objectives and tactics (activities) constitute the tactical statements.

PRE-PLANNING

Although a formal pre-planning stage is not provided, it is suggested that, in the absence of a strategic plan, a management questionnaire, an outline for which is given, should be distributed to relevant personnel and the results analysed using a technique termed goal or critical success factor analysis. One of its objectives is to plan the best approach to be used for a strategic planning study.

STRATEGIC IMPLEMENTATION AND MANAGEMENT

Strategic implementation

This begins with the distribution of the strategic plan, together with the management questionnaire mentioned above, to individuals who were not involved in the planning process. Then follow the analysis and design phases of Information Engineering (see Chapter 10). These build on the strategic plan and produce *strategic* and *tactical* models, which schematically represent the organization in terms of data and information needed for management. The models are termed data maps, which are normalized entity model diagrams, and they also contain the objectives determined earlier in the planning study.

Strategic management

This stage consists of using the strategic model (obtained from the later analysis phase) as a basis for testing strategic alternatives, and it is claimed that the model facilitates an estimation of the effects of these alternatives. An advantage is that feedback as to positive or negative effects may be obtained more quickly compared to trying out the strategy directly in the organization. Refinements may then be made to the organization's plans, and the use of this approach, in a more or less continuous strategic planning process, is held to be a desirable management tool for coping with business changes and quickly estimating the effects of strategic alternatives.

Dickson and Wetherbe approach

INTRODUCTION

The broad aims of strategic planning (Dickson and Wetherbe, 1985) are to establish potentially beneficial management information system (MIS) services, to perform some cost/benefit analysis and to allocate resources within the potential project portfolio. However, they note that some problems with strategic planning are:

1. Alignment of MIS plan with organizational plan
2. Design of an IS architecture for use as a base to develop applications
3. Allocation of IS development and operations among competing applications
4. Selection and use of methods for planning

APPROACH STAGES

The four-stage approach is shown in Fig. 14.10.

Strategic MIS planning

The aim of this stage is to establish the relationship between the overall organizational plan and the MIS plan.

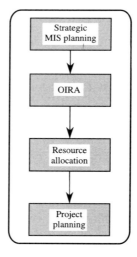

Figure 14.10 Stages of the Dickson and Wetherbe planning approach

- *Assess organizational objectives and strategies.* This step reviews the strategic plan of the organization and identifies major claimant groups and their objectives.
- *Set MIS mission.* This defines a new or revised set of MIS objectives.
- *Assess the organizational environment.* The factors considered here are the current business environment and new opportunities; the current applications portfolio, current MIS capabilities and new technology; and the MIS image, stage of MIS maturity and MIS personal skills assessment.
- *Set MIS policies, objectives and strategies.* This step will set the organization structures, resource allocation mechanisms, management processes and objectives, as well as setting the technology focus. It is suggested that the strategy set transformation approach (King, 1978) may be used to assist in this phase.

  Problems may occur in establishing organizational goals and priorities. These may be unwritten or they may be set in terms not useful for IS planning. Basing future systems on user proposals may result in unsuitable systems.

Organizational information requirements analysis (OIRA)

The aim here is to identify broad organizational information requirements and to establish a strategic information architecture. This can be used as a base for specific application development projects. Two steps are distinguished. In the first step, high-level current and projected information needs to support organizational activities are obtained. Secondly, a development plan is assembled from the information architecture which defines specific IS projects, project priority and the project development plan. Techniques suggested include BSP (IBM, 1984), BIAIT (Burnstine, 1980) and critical success factors (Rockart and Flannery, 1983).

The two problems found here are the eliciting of a correct set of information needs and the selection of an architecture (assuming complete and correct information needs understanding) based on those requirements. A vague set of information needs will give rise to a very large set of alternatives for the architecture. It may be difficult for managers unaware of computer potential to consider computer-based solutions.

Resource allocation

This step develops hardware, software, communications, facilities, personnel and financial plans needed to implement the plan from the previous OIRA stage. Specific techniques suggested to assist with this are mainly concerned with financial aspects, either from prioritizing projects in terms of return on investment or for MIS planning and control.

A rationally based allocation of resources is always difficult, as the advantages of different alternatives cannot always be quantified. In addition, future systems may not fit into current organizational structures, which would help to establish priorities.

Project planning

Project planning produces a development plan for all future IS projects and evaluates projects for requirements and difficulties, defines necessary activities, estimates resources (time, money) required and sets completion dates. Suggested techniques are PERT charts (showing project tasks and their task dependencies), milestone planning (establishing checkpoints without prior resource commitment) and Gantt charts (showing tasks, with start and end dates).

PRE-PLANNING

It may not be necessary to execute all stages of the above approach. For example, if major strategies do not change, an annual cycle beginning with OIRA may be adequate. The authors provide a 'stage assessment' guide which may be used to determine which stage or stages are needed in an organization.

Discussion

APPROACH COMPARISON

Table 14.3 shows the main features of the three approaches, in terms of their processes, and Table 14.4 shows their products. From Table 14.3, it can be seen that the Dickson and Wetherbe approach appears to be the most comprehensive, as it is the broadest of the approaches, extending to individual project planning. However, when the depth of the approaches is compared, SISP is easily the most thorough. Information Engineering only covers organizational strategies, objectives and activities, and the information architecture is built in fact by the next, analysis, stage of Information Engineering. Table 14.4 also shows that the Dickson and Wetherbe approach produces a wider range of products, but the level of detail is much less than that given by the SISP approach. For example, the information architecture produced by the Dickson and Wetherbe approach is expressed in terms of undefined information categories, and entity modelling is not considered. Figure 14.11 compares the three approaches in terms of breadth and depth.

Another feature of interest is that the Dickson and Wetherbe and SISP approaches are top-down, starting from organizational objectives, while the formal planning approach of Information Engineering is bottom-up, starting

Figure 14.12 attempts to illustrate this. Analysis begins by accumulating a mass of information concerning the organization and its environment. A set of candidate activities is obtained by comparing objectives with current activities, and awareness of any problems may help to focus attention on key areas.

However, these are ideal activities. Applying the realism of SWOT (strengths, weaknesses, opportunities and threats) to the organization and the environment then reduces this set. Looking for technological opportunity may further modify the set to produce the final key set of selected activities and information needs. The analysis phase thus consists of an expansion followed by a contraction. The design phase, using these activities and information needs as a base, then produces the detailed plans for their implementation.

ARE PROBLEMS SOLVED BY STRATEGIC PLANNING?

We may legitimately ask the question: does strategic planning provide solutions to the three problems mentioned at the beginning of this chapter? It is possibly too early in the life of strategic planning as a process to tell, as the point is made (Galliers, 1989) that most of the evidence suggesting that solutions are provided is based on personal experience and case studies, and that unsuccessful planning exercises do occur; however, Earl (1987) agrees that there is much research to be done. It is difficult, for example, to monitor the effects of a plan or to evaluate plan effectiveness, owing to the complexity and time-scales involved.

It is clear that, properly done, planning will spend a considerable amount of time in establishing required organizational activities and their information needs. This should help to make later applications, developed within such a context, fit their requirements more closely than before. It should be realized that this does not come free and that the early stages of development must now take longer than before, if a correct information architecture for the required activities is to be constructed.

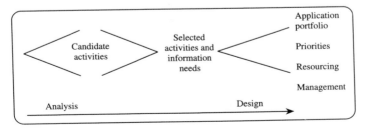

Figure 14.12   Analysis and design phases in strategic planning

CRITICISMS OF STRATEGIC PLANNING

A criticism may be made that planning is merely a high-level way to fix requirements, which is an unrealistic endeavour over a long time-scale. The dangers of a five-year plan in a rapidly changing business environment are obvious, as systems may be delivered that match the plan but are no longer useful.

A second objection is that organizational objectives are not agreed or known, but are subject to continual negotiation. Therefore, to attempt to discover organizational objectives, strategies and policies may at best only give a view at a particular moment in time and at worst will produce a set of compromises that will be interpreted in different ways by different key individuals, when systems reflecting the objectives are actually designed and implemented.

It is also suggested (Galliers, 1989) that different perceptions of the success of the planning process outcome exist, as senior managers emphasize the importance of increased user involvement, among other factors, whereas IT planners rank a clearer view of overall IS requirements more highly. The fact that many planning approaches provide only one basic method, with no guidance for organizations to tailor this to their particular requirements, is criticized (Hirschheim, 1989), and Hirschheim puts forward a framework of four different types of planning. The framework is based on whether planning should, or should not, be undertaken as part of the organization's long-range planning and whether the plan should be oriented around the information needs of managers or the benefits to the organization to be gained by applying information technology.

Finally, there are reports that question whether planning can bring about competitive advantage. A study by the Kobler Unit of Imperial College, London (Hochstrasser and Griffiths, 1990), claims that most users find it difficult to acquire a competitive advantage through IT, that the few successes that have been achieved are accidental and that these successes will be difficult to emulate, and also casts doubt upon the notion that acquiring a competitive advantage is a valid objective in the long term.

Summary

We have introduced the topic of strategic planning in this chapter and we stated that it addressed the first quality problem in systems development. We began by outlining the content of a strategic plan and then discussed, in general, three reasons why individuals and organizations find it useful to plan. The problems that strategic planning implicitly address were then presented, together with solutions. The topic was then defined, as well as its objectives, and then the three-eras model was briefly discussed, describing how the different types of information system being produced today require a planning context to be effective.

A general model of the strategic planning process was then presented, consisting of two phases, pre-planning and planning. The process was described in terms of inputs, outputs, models and techniques. Three approaches to strategic planning were then described. The SISP approach consisted of two phases, planning for planning and planning. Planning for planning was discussed and then the four stages of planning — internal and external business environments and internal and external IS/IT environments — were described. We then illustrated the SISP approach with a case study, based on the strategic plan of a local authority.

The two methods of Information Engineering were then described, consisting of formal and informal strategic planning. These were intended to be used in different contexts. Finally, the four-stage Dickson and Wetherbe approach was briefly introduced.

The approaches were then compared, in terms of breadth of processes, products and combined breadth and depth. An overview model was briefly developed and the chapter concluded by asking whether strategic planning solves the problems it addresses and considering several criticisms.

Discussion questions

1. Consider the plans that you may have made for yourself, perhaps when planning to write a piece of software. Do you agree with the reasons for planning given at the beginning of the chapter? If not, why not?
2. Identify a transaction processing system (TPS) in an organization with which you are familiar and compare it with a system used for providing information to managers for decision making. Why should planning be less necessary for the TPS than for the decision-making system?
3. Would you agree with the Information Engineering approach that formal strategic planning should require more user time than informal planning?
4. What would you say were the difficulties in carrying out the strategic management activity of Information Engineering? What exactly does the 'strategic model' that is being used consist of?
5. How would you begin to improve the OIRA phase of the Dickson and Wetherbe approach?
6. Can you think of a good way to determine what kind of strategic planning exercise might be required in an organization?
7. How would you begin to plan for the evaluation in an organization of (a) the strategic planning process, (b) the effects of strategic planning?
8. Bearing in mind the problems caused by the hard approach assumptions discussed in Chapter 12, do you think the information analysis techniques described in the strategic planning approaches will be successful?

References

Burnstine, D. C. (1980) *BIAIT: An Emerging Management Discipline*, BIAIT International, New York.

Checkland, P. B. (1981) *Systems Thinking, Systems Practice*, Wiley, London.

Dickson, G. W. and J. C. Wetherbe, (1985) *The Management of Information Systems*, McGraw-Hill, New York.

Earl, M. J. (1987) 'Information systems strategy formulation' in *Critical Issues in Information Systems Research*, R. A. Boland and R. A. Hirschheim (eds), Wiley, Chichester, pp. 157–178.

Fairbairn, D. (1989) 'Senior management's perception of information technology and information systems' function' in *Information Management and Planning: Database 87*, P. Feldman, L. Bhabuta and S. Holloway (eds), Gower Technical, Aldershot, pp. 27–32.

Fawthrop, A. (1990) 'The systems engineering approach within British Telecom' in *SE 90 (Proceedings of Software Engineering 90, Brighton, July 1990)*, P. A. V. Hall (ed.), Cambridge University Press, Cambridge, pp. 514–528.

Finkelstein, C. (1989) *An Introduction to Information Engineering*, Addison-Wesley, Wokingham.

Galliers, R. D. (1989) 'Applied research in information systems planning' in *Information Management and Planning: Database 87*, P. Feldman, L. Bhabuta and S. Holloway (eds), Gower Technical, Aldershot, pp. 45–58.

Hirschheim, R. A. (1989) 'Information management planning: an implementation perspective' in *Information Management and Planning: Database 87*, P. Feldman, L. Bhabuta and S. Holloway (eds), Gower Technical, Aldershot, pp. 1–15.

Hochstrasser, B. and C. Griffiths (1990) *Regaining Control of IT Investments: A Handbook for Senior UK Management*, Kobler Unit, Imperial College, 180 Queens Gate, London SW7 2BZ.

IBM Corporation (1984) *Business Systems Planning: Information Systems Planning Guide*, GE20-0527-04. Available from IBM Technical Publications Centre, PO Box 117, Basingstoke RG21 1EJ.

King, W. R. (1978) 'Strategic planning for management information systems', *MIS Quarterly*, vol. 2, no. 1, pp. 27–37.

Olle, T. W., J. Hagelstein, I. G. MacDonald, C. Rolland, H. G. Sol, F. J. M. Van Assche, A. A. Verrijn-Stuart (1988) *Information Systems Methodologies: A Framework for Understanding*, Addison-Wesley, Wokingham.

Porter, M. E. (1987) *Competitive Advantage: Creating and Sustaining Superior Performance*, The Free Press, New York.

Rockart, J. F. and L. S. Flannery (1983) 'The management of end user computing', *Communications of the ACM*, vol. 26, no. 10, pp. 776–784.

Ward, J., P. Griffiths and P. Whitmore (1990) *Strategic Planning for Information Systems*, Wiley, Chichester.

15
Social issues and information systems

Introduction

Previous chapters in part four have looked at information systems mostly from the viewpoint of the considerations needed to be taken into account in system planning and design by social, psychological and economic factors within the organization. In contrast, this chapter will discuss issues raised by the effects of information systems on individuals, organizations and society, covering the security and reliability problems discussed briefly in Chapter 1.

The first section is concerned with the social effects of information systems on organizations, considering several evolutionary models of organizations, followed by a brief discussion of the implications for centralization and organization size of increasing investment in information systems. The notion of software as a resource is then considered, where software has its value estimated and is treated as any other organizational asset. The next section briefly discusses safety issues arising from system reliability problems.

Legal issues are then considered, emphasizing security problems, and the Data Protection Act 1984 is discussed, with its implications for privacy of personal data and organizational behaviour with respect to data security. Two further aspects of security are examined — human threats such as hacking, involving a discussion of the Computer Misuse Act 1990, and software threats such as viruses.

Information systems evolution

In this section, we look at the effect of information systems on the organization as a whole. To do this a number of evolutionary models are presented; they discuss the stages through which organizations pass as computerization increases.

NOLAN MODEL

The Nolan model (Nolan, 1979) was developed in the 1970s by empirical research in a number of large US organizations, and describes six stages through which an organization evolves as it spends an increasing amount of money on computerization of its activities. The six stages are:

1. *Initiation*. The emphasis in the first stage is on computerizing functional activities, with the aim of cost reduction.
2. *Contagion*. Users now have overcome their unfamiliarity with the computer and the demand for new applications rapidly increases. Computer services may be free to users. The IT department expands to meet demand in a haphazard fashion with little planning or control.
3. *Control*. Senior management move to control expansion by limiting or cutting the IT budget. Costs and benefits of systems are made more visible, by installing management techniques in the IT department. Emphasis is placed on methods and documentation.
4. *Integration*. Existing systems are integrated usually via database technology. User accountability for their systems becomes accepted as a principle and the function of the IT department is to service users.
5. *Data administration*. Database experience has been gained and a data-oriented approach to systems development is adopted. This consists of a data administration function that controls organizational data, allied to an emphasis on common data for planning new, integrated systems. Users are responsible for the appropriate use of the technology.
6. *Maturity*. In the last stage, the technology is completely integrated with the organizational objectives, and there is joint user and IT department accountability for IT resources.

The model may be used in several ways. For example, information about many organizations may be gathered and used to determine the approximate degree of maturity that has been reached. Within an organization, the model may be used to determine the degree of maturity of different aspects of the organization, such as its extent of planning or user participation, which is a useful starting point for planning new systems, giving clues about the degree of staff expertise or the level of computer awareness in a user department.

The model has several weaknesses, one of which is that it has not been verified by later research. Examination of the model suggests that, particularly for the first two stages, it is a historic model only. In addition, there is some confusion about accountability in the last three stages.

HIRSCHHEIM MODEL

A three-stage model is presented which is claimed to overcome some of the deficiencies of the Nolan model (Hirschheim *et al.*, 1988). The stages are:

1. *Delivery.* The focus in the IT department is inward, concentrating on technical problems and issues with hardware and software. It is important to achieve management credibility by delivering a product (not necessarily what users want) within budget and on time.
2. *Reorientation.* It now becomes important to forge good relationships with user departments, as systems are used in more and more parts of the organization. The emphasis is on delivering what users want.
3. *Reorganization.* User departments have become aware of the benefits that can be obtained from successful systems and top management moves to planning for integration of organizational and technological issues.

This model is very much an intuitive model, as it is not based on empirical research, as is the Nolan model, and it is a two-step model to achieving planning.

THREE-ERA MODEL

This model has already been briefly mentioned in the chapter on strategic planning (Ward, Griffiths and Whitmore, 1990). It focuses on only one aspect of the organization and its use of computers, which is the broad type of activity supported. The model is:

1. *Data processing (DP) era.* Most information systems only automated the operational activities of organizations. The aims were cost reduction and productivity gains.
2. *Management information systems (MIS) era.* This era emphasized user-initiated enquiry systems and information analysis for assisting management activities. Technologically, these management systems shared a common base of information with DP systems. The aim was to enhance management effectiveness or to improve the quality of product or of service.
3. *Strategic information systems (SIS) era.* The current era is characterized by the aims of improving organizational competitiveness by changing the ways in which objectives are achieved, or by changing objectives. New types of system may incorporate more intelligence than before or they may process a wider variety of information than traditional data.

The authors add that, broadly, the DP era dates from the 1960s, the MIS era from the 1970s and the SIS era from the 1980s. However it is possible for different computerized parts of an organization to be in different eras.

SUMMARY OF EVOLUTIONARY MODELS

The models described above are not claimed to be very precise, but they may be useful tools for applying to different departments or functions within an organization, to determine the rate at which computerization has proceeded.

For example, it is common to find that the accounting function is ahead of departments such as production, in terms of computerization. Different rates of computerization can obviously affect the potential for integration.

Information systems and organizational structure

There are few studies of how the use of information technology affects organizations as a whole, but in Gurbaxani and Whang (1991) this topic is developed, in a theoretical manner, concentrating on two aspects of organizational structure: the degree of centralization and the organizational size.

Two kinds of costs are outlined, termed internal and external co-ordination costs. Internal co-ordination costs are those that are incurred in the endeavour to keep the behaviour of organizational employees in line with the objectives set by senior management, shareholders and so on. For highly centralized organizations, decision information costs are high, as information is more easily and quickly accessible to employees at the bottom of the organizational pyramid, and it therefore has to be channelled upwards to management for them to make decisions.

For highly decentralized organizations, what are termed agency costs are high, as the organizational leaders have to constantly monitor employee behaviour to ensure that they adhere to objectives. The aim is thus to minimize both decision information costs and agency costs.

For organizational size, external co-ordination costs are important. These are costs incurred when carrying out transactions in the external market, such as transactions with suppliers or customers. Costs are for writing and enforcing contracts, searching for information, transportation and so on. However, there may come a point when costs may be reduced by performing the transactions within the organization, and this may be achieved by mergers of the vertical integration type, where an organization acquires, for example, a key supplier.

The conclusions of the study are that information technology does not have a decisive effect on the degree of centralization, as both decision information and agency costs may be reduced. Other factors, such as those discussed in Chapter 13, are important for deciding the degree of centralization or whether a hybrid (mixed) structure is appropriate.

For organizational size, the authors find that if information technology is reducing internal co-ordination costs, there is a trend, other factors being equal, to larger organizations, measured both vertically (the range of value chain the organization spans) and horizontally (the number and corresponding share of different markets in which the organization sells its goods or services). This is also due to economies of scale being made possible by new technology.

However, organizations must balance their internal co-ordination costs with their external co-ordination costs, because in order to grow larger, reducing external costs, internal costs may be increased, so possibly deriving no overall benefit. Some organizations therefore may choose not to grow larger and may instead establish close contacts with customer or supplier organizations, if they do not wish to incur the larger internal costs of a larger organization. Depending on current cost structures, as well as other factors, organizations may grow larger or smaller.

Software as a resource

Although development budgets often run into millions of pounds and the software produced is maintained at even greater cost, it is not current accounting practice to show software as an asset on the organization balance sheet. This means that software is not audited, costed or depreciated, as are other assets, but, more importantly, it is not budgeted to be replaced when it is of no use. However, in an interesting paper, Rigby and Norris (1990) describe a project that involved them in developing a cost model for the software owned by British Telecom in the United Kingdom.

The basic idea underlying the approach is that of the 'software death cycle', as opposed to the traditional life-cycle concept. The emphasis is thus on the cost of keeping an existing system productive, rather than on the cost of developing a new system.

The steps suggested are:

1. *Audit software.* This consists of identifying items such as which programs and in what computers and locations they are held, program support environment (for example, operating system environment), associated documentation, licensed (how many licences) or developed software, and current or disused software. An advantage from this audit is that software ownership can be established, which is useful for assigning costs later.
2. *Assess value.* It is obviously difficult to put an objective value on an item of software, and the factors that should be considered range from financial to goodwill factors. Two extremes are suggested. Firstly, there is commercial software that has been bought to do a job or that has been developed to be sold, and its value can be partially estimated by determining whether or not it earns money. However, the costs, particularly of developed software, such as development costs or maintenance costs, are not so easy to determine. At the other extreme, there is strategic software, which performs an essential function within the organization. The value of this is often unrelated to the actual cost of development or potential market value, and is usually highly subjective.

After a value has been assigned, it should be possible to draw a model of income/cost over time, and an actual situation, relating to a program product that was developed and sold commercially, is shown in Fig. 15.1. Ideally, when the costs outweigh the benefits the software should be retired.

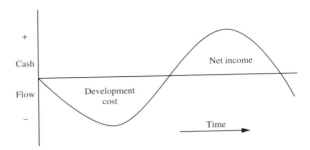

Figure 15.1 Cash flow model of software ownership

The figure shows that the initial costs are development costs (time and money taken to develop the software with no income) and the net income is income from sales minus the cost of support for the product.

3. *Establish value system.* Once the value has been assessed, a planning system for the software is suggested. This is done using two key variables — the present and predicted values of the software — to show how the software currently fits onto a planning grid based on the business portfolio analysis model, often termed the 'Boston matrix'. This is shown in Fig. 15.2 and is basically another way of representing the information in Fig. 15.1.

A product at the beginning of the curve in Fig. 15.1 has only a low present value but a high predicted value, and is thus a 'wild cat', which has valuable potential. A product that has crossed from negative to positive income is a 'star', and should be strongly supported. 'Cash cows' graze around the top of the curve and represent money in the bank, while 'dogs' are going downhill with no long-term value and should be removed.

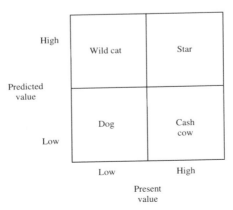

Figure 15.2 Boston matrix for assessing software value

It should be noted that a problem with this simple model is that the predicted value of the software relies heavily upon business judgement, and no method is given for taking subjective views of the current value of the software into account.

Reliability

The problem of system reliability is extremely important, often involving life or death or the safety of large financial transactions. Concern is increasing that, as software begins to play a larger part in these systems with an attendant reduction in human supervision, the chance of system failure is increasing. The consequences of such failure for human life, confidence in financial institutions or markets, or the environment, for example, may be far-reaching.

Reliability is concerned with the correctness of an implementation with respect to a specification. For example, there should be no bugs in code written from a specification. Typical environments in which problems may occur are: space flight, missile and air defence, naval weapon systems, military and commercial aviation, surface transport (rail and bus), cars, robots (at least six cases of robot-induced death have occurred), financial systems and medicine.

For example, there have been reports of serious failure in safety-critical software which controls instrument panels on BA planes (*The Guardian*, 8 September 1990) and signal centres for British Rail (*The Guardian*, 23 July 1990). Another example concerns the loss of the Phobos I spacecraft (quoted in *Software Engineering News*, October 1988), where a mistaken 'commit suicide' command caused the spacecraft's solar panels to point the wrong way, preventing the batteries from charging, and ultimately causing a loss of power to the spacecraft.

As discussed in Chapter 6, a current approach to this problem is to attempt to prove software (and hardware) correct, with the use of formal methods.

Data Protection Act

Thailand — a constitutional monarchy with a parliament largely dominated by the military — has taken the Orwellian step that most Western democracies have been afraid to take. The Thai government this month inaugurated a centralized database system to track and to cross-reference vital information on each of its 55 million citizens. The system includes a Population Identification Number (PIN) with a required computer-readable ID card with photo, thumbprint, and embedded personal data. The system will store data of birth, ancestral history, and family make-up and was designed to track voting patterns, domestic and foreign travel, and social welfare. Eventually 12 000 users, including law enforcement, will have access by network terminals. It is the largest government relational database system in the world. 'The people feel that the system will protect them,' says the director of the Central Population Database Center in Bangkok.

ACM *Software Engineering News*, October 1990

INTRODUCTION

The Data Protection Act 1984 is concerned with the privacy and security of certain types of 'automatically processed information' — broadly, data held in computer systems. The main aim of the Act is to protect the information rights of the individual, by restricting the availability of sensitive personal data, such as data concerning the individual's health, as well as allowing access rights to the individual about whom data is recorded. In addition, the Act allows data to be transferred freely, for purposes such as trade, between the United Kingdom and other European countries that have ratified the Council of Europe Convention on Data Protection.

The Act only applies to automatically processed information; it does not cover data that is only processed manually. Legal requirements are set out concerning the registration availability and transfer of what is defined as 'personal data'. This is information recorded on a computer about living, identifiable *individuals*, and not, therefore, organizations. Information includes opinions as well as facts.

The 'data user' is the user (often an organization) of the data who is legally responsible for data acquisition and who controls the content and organization of the data. The 'data subject' is the individual to whom personal data relates. Obligations are placed on data users, as well as 'computer bureaux', who are people or organizations who broadly process personal data for data users.

The Act therefore has implications for those who design and operate information systems that contain personal data.

REGISTRATION

The data user is advised to appoint a Data Protection Officer, who is responsible for properly registering personal data and describing the purpose for which it is to be used and who is the person to whom an individual would make a subject access request concerning his or her personal data. It is a criminal offence not to register such data. It is also a criminal offence if a registered data user knowingly or recklessly operates outside the descriptions contained in their register entries.

Registration of personal data and its use is accomplished by applying for a register entry with the Data Protection Registrar, who is an independent official reporting directly to Parliament. The Registrar's duties include:

1. Establishing the Register of Data Users and Computer Bureaux and making it publicly available.
2. Considering complaints about alleged breaches of the Act and, where appropriate, prosecuting offenders or serving notices on registered data users or computer bureaux.
3. Encouraging the development of codes of practice to assist with compliance to the eight principles, discussed below.

EXEMPTION FROM REGISTRATION

Some personal data is exempt from registration under the Act. Such data need not be registered and the Registrar has no power over the data. The exemptions are:

1. Data held by an individual concerning his or her personal, family or household affairs.
2. Information that the law requires to be made public, such as the electoral register maintained by an electoral registration officer.
3. Data to safeguard national security.
4. Data for payroll, pensions and accounts purposes, but the data may be held only for certain specified purposes and there are restrictions on how it is made available to others.
5. Data concerning unincorporated members' clubs.
6. Data for mailing lists. The data must be held only for distribution purposes and there are four conditions that must be met.

PRINCIPLES

There are eight principles that should be used as a guide when developing and operating information systems. They are intended to protect the rights of the individuals about whom personal data is recorded. All principles apply to data users, but only the eighth principle applies to computer bureaux.

First principle

The information to be contained in personal data shall be obtained, and personal data shall be processed, fairly and lawfully.

This means that the person who provides information should not be deceived or misled with regard to the purpose for which the data will be used or disclosed.

Second principle

Personal data shall be held only for one or more specified and lawful purposes.

To comply with this principle it is necessary to register under the Act and process data in accordance with the purpose registered.

Third principle

Personal data held for any purpose or purposes shall not be used or disclosed in any manner incompatible with that purpose or those purposes.

The use or disclosure of data must, for compliance with this principle, only be in accordance with the registered purpose.

Fourth principle

Personal data held for any purpose or purposes shall be adequate, relevant and not excessive in relation to that purpose or those purposes.

The intent of this principle is that the data held shall be the minimum that is necessary and sufficient for its purpose.

Fifth principle

Personal data shall be accurate and, where necessary, kept up to date.

Data is only considered to be inaccurate within this principle if it is incorrect or misleading with regard to a fact, as opposed to an opinion.

Sixth principle

Personal data held for any purpose or purposes shall not be kept for longer than is necessary for that purpose or those purposes.

This concerns the length of time that data may reasonably be held for the registered purpose. Personal data held for historical, statistical or research purposes may be kept indefinitely, as long as the data is not used so as to cause damage or distress to any data subject.

Seventh principle

An individual shall be entitled:
(a) at reasonable intervals and without undue delay or expense —
    (i) to be informed by any data user whether they hold personal data of which that individual is the subject; and

(ii) to access any such data held by a data user; and
(b) where appropriate, to have such data corrected or erased.

Section 21 of the Act and Registrar's Guideline Number 5 set out the rights of subject access in more detail. 'Reasonable intervals' between access requests will depend upon the type of data and how frequently it is updated. 'Appropriate' in relation to data correction or erasure is concerned with compliance with the other principles.

Eighth principle

Appropriate security measures shall be taken against unauthorised access to, or alteration, disclosure or destruction of, personal data and against accidental loss or destruction of personal data.

The issue of security applies here, to both data users and computer bureaux, and covers:

1. Unauthorized access, either physically into the computer premises or into the hardware or software.
2. Alteration, disclosure or destruction of data without the authorization of the data user.
3. Accidental loss or destruction of data because of unreliable hardware, software or back-up and recovery procedures.

Particular consideration should be given to security needs in the light of the sensitivity of the data and the harm that could result from any of the above breaches. Consideration should also be paid to the physical security of the computer installation, to software security measures and to the reliability of staff accessing data. This might include staff selection on the basis of competence, as well as sound induction and adequate system training.

SUBJECT ACCESS

Access rights

An important part of the Act relates to the access rights of data subjects, that is those individuals about whom personal data is recorded. The data subject has the responsibility to apply for 'subject access', and must specify the registered entry under which access is being requested. This is because data users may register a number of purposes under different subject entries.

All information, current and historical, held under a registered entry about the individual should be provided. The data user must respond to a subject access request within *40 days* of having received the request. The period starts when the data user can identify the data subject and locate the relevant data. The data user may charge a fee for access to each registered entry, and there is a statutory maximum.

Exemptions from subject access

Data held for certain purposes is exempt from subject access. For example:

1. Data held for the purposes of crime prevention, or tax or duty assessment or collection.
2. Government department data concerned with the making of judicial appointments.
3. Data consisting of information that might form the basis of a claim of legal professional privilege, that is concerning lawyer–client confidentiality.
4. Data held only for the preparation of statistics or the carrying out of research.
5. Back-up data, that is, data held only for replacing other data if it is lost.
6. Data held by credit reference agencies. This is available under the Consumer Credit Act 1974.
7. Data that incriminates the data user.
8. Data exempt from registration.

In addition, the relevant Secretary of State may, with an order, exempt certain personal data from subject access. Currently, the following orders have been made:

- *Health.* 'Health data' is defined as data relating to the mental or physical health of the data subject. Furthermore, the data must have been collected by or on behalf of a health professional, such as doctors, dentists, chemists, nurses, opticians and clinical psychologists. Data users may be, for example, employers, insurance companies, health authorities or GPs. Subject access rights are removed either where subject access would be likely to cause serious harm to the health of the data subject or where another individual would be identified. Before data users take a decision on a subject access request they must consult the 'appropriate health professional', who might be, for example, the most recent GP of the data subject.
- *Social work.* Access rights of data subjects are removed where disclosure would either cause serious harm or where another individual is identified. Data users here are primarily health and local authorities, or institutions concerned with care such as the NSPCC. As for health data, subject access rights are removed either where subject access would be likely to cause serious harm to the health of the data subject or where another individual would be identified. There is no requirement for another professional to be consulted before disclosure or non-disclosure.

technology. The plan usually refers to organizational aims and the information systems necessary to support those aims, together with a description of the management structure and technology that will be required to implement the plan.

Strategy this describes how an organization will achieve its (long-term) objectives.

System an assembly of related parts that act together as a whole; something that may be decomposed into the following components: input, output, process, boundary and environment.

Systems development process the process of developing an information system starting from understanding user requirements, through implementing a system that meets those requirements, to maintaining that system when changes occur.

System life cycle another term for the systems development process, intended to emphasize the fact that, taking the maintenance phase into account, a system effectively goes through many cycles of analysis, design and implementation.

Tool an automated aid to an activity within the systems development process.

Transaction a self-contained event that is a fundamental unit of the main area of concern of the organization. For example, a bank withdrawal or a concert seat reservation.

Transaction processing system (TPS) this is a system that provides procedures to record and make available information concerning the occurrence of transactions (see above) in the application.

Usability a property of an information system such that the users can work with the system and use it as intended. This is usually interpreted as relating to the design and use of the human computer interface.

User requirement this is the user perception of a system that the user wants, interacting with the organization, together with the expected benefits that the user hopes will materialize from the use of the system. It is the set of functional and non-functional requirements that the user has of a system and its interaction with the user organization.

Validation the activity that checks that a phase product of the systems development process is correct in terms of the user requirement. The term usually implies that users are involved in checking that the phase product has captured the requirement correctly.

Verification the activity that checks that a phase product is internally (syntactically) correct. It is usually performed by designers. For example, checking for a DFD that all data flows have names, or that no flows exist between data stores.

Author Index

Index